For Carol –

WILD SPECTACLE

Seeking Wonders in a World beyond Humans

JANISSE RAY

Janisse Ray

11/19/2021
Apalachicola, FL

TRINITY UNIVERSITY PRESS
San Antonio, Texas

Trinity University Press
San Antonio, Texas 78212

Copyright © 2021 by Janisse Ray

Book design by BookMatters, Berkeley
Jacket design by Derek Thornton, Notch Design
Jacket art, Bridgeman Images 235633
Author photo by Christopher Ian Smith

ISBN 978-1-59534-957-6 hardcover
ISBN 978-1-59534-958-3 ebook

Trinity University Press strives to produce its books using methods and
materials in an environmentally sensitive manner. We favor working with
manufacturers that practice sustainable management of all natural
resources, produce paper using recycled stock, and manage forests with
the best possible practices for people, biodiversity, and sustainability. The
press is a member of the Green Press Initiative, a nonprofit program
dedicated to supporting publishers in their efforts to reduce their impacts
on endangered forests, climate change, and forest-dependent communities.

The paper used in this publication meets the minimum requirements of
the American National Standard for Information Sciences—Permanence
of Paper for Printed Library Materials, ANSI 39.48-1992.

CIP data on file at the Library of Congress

25 24 23 22 21 | 5 4 3 2 1

To my teachers,
past, present, and future

CONTENTS

PREFACE

Out beyond houses and mailboxes, roads and bridges, a person can see a realm that exists alongside this world in which we humans live.

I say again, another world flanks the constructed one. Often the view from ours is skewed, as through fractile glass, limited by narrow apertures of scope and angle and crack, the view fleeting. We can't see it on demand.

In the wild world, relationship is evolutionary, time is geologic, beauty is intelligent. There we find ourselves under a powerful spell.

Although I was reared on a junkyard by parents who did not waste time hiking or camping, I knew pine trees and pitcher plants, bobcats and brown thrashers, as my people. I understood wild things as beings with intentions, foremost a searing desire to live pleasant, fulfilling lives.

Once the storyteller Joseph Bruchac explained to me that there are people to whom animals are attracted, to whom animals listen. Later I met such a person, an Abenaki man named O'annes. He visited environmental studies classes at a university where I was in residence, and my colleagues described an odd thing that often occurred during O'annes's visits. As he sat outside, on a green or by a lake, talking to students about his ethos, an animal would ease up to listen. It might be a heron or squirrel, it might be an alligator. When O'annes lectured my own class, we convened outside, and I was gobsmacked when a black racer came sliding along with its head lifted from the mown grass, startling my students. It circled behind O'annes before hunkering down in shrubbery, as if to eavesdrop.

The essays in this book are about a desire to immerse myself in the varied wild, to survey the territory of wildness, to be wild, and, perhaps, to become the kind of person who listens to animals and to whom animals listen. I explore places of natural spectacle and abundance, the less mitigated and trammeled the better. Because I was born in the twentieth century, I have missed many wonders that mavericks like Bartram, Carver, Crazy Horse, Muir, Sacagawea, and Tubman (indeed, anyone able to notice such things) saw—flocks of passenger pigeons, packs of red wolves, sleuths of bears. I mourn that loss. On the other hand, I have seen wonders that inhabitants decades hence will be unable to see.

I have, in my luckiest moments, lived heart-pounding flashes of wild spectacle.

Many of the essays in this book, collected from a long span of years, involve travel. I am a traveler, but as the climate destabilized, I increasingly could not justify it, another form of personal consumption. Over a decade ago, for the sake of the climate and life on earth, I quit flying altogether and dramatically decreased my use of fossil fuels. I still travel, and the travel is more local, more interior, as well as more occasional. I am deeply grateful for the natural resources, especially the liquefied ancient plants compressed by the earth that, adapted to combustion engines, allowed me access to some of the wonders I've seen.

This book is about a longing to experience our landscapes deeply and grandly—as dwellers, as eaters, as seekers, as learners, and especially as passers-through, because in the grandest scheme we are all visitors, just visiting this planet, death the trackless wilderness to be explored. Here is what I found, what I saw, what I heard, what I thought, and what I learned when I sojourned in the wild.

PART I   MERIDIAN

Exaltation of Elk

IN THE BACKCOUNTRY OF THE WEST
Bob Marshall Wilderness, Montana

Light rain had fallen all afternoon, and the sky looked as if it planned on raining into the night. I had been walking for five days through Montana wilderness with my husband, Raven, a mail carrier becoming a painter. Since early morning we had traversed ten miles. Night had not yet arrived, although the afternoon was so dim it hardly seemed day.

We were wearing cheap rainsuits and were walking in plastic bags, between our socks and boots. Our packs were covered with trash bags. Raven's dark hair curled from his ballcap, and drops of mist caught in his long eyelashes. In the vast, rainy wilderness of the Bob Marshall, we had become so small we were almost invisible.

We crossed by pack bridge over the North Fork of the Sun River and entered the territory of a terrible fire some years past, wherein a tremendous amount of biomass had converted to charcoal and ash. The strange gray landscape was a graveyard of trees. Trees stood dead or lay where they had fallen, crossing and crisscrossing on the windy slopes like toothpicks. Here and there young firs had begun a slow scramble toward recovery.

We stopped to rest on a flowery bluff at the junction of Headquarters and Dryden Creek Trails. Indian paintbrush and silvery lupine swept color through a canvas of grasses underfoot while below, Headquarters Creek rushed along through a stony bed. Wilderness stretched for many miles on all sides, and fog crept in among the blackened sticks of trees.

Across the creek Beartop Mountain rose. On top of that mountain was a lookout cabin, we knew, at 8,000 feet, and a lookie would be in it now, reading and glancing out his big glass windows. We couldn't see the cabin from the bluff, although earlier we had seen smoke drifting from its chimney.

We were invisible to the lookie. Being invisible made us quiet, and that made us more invisible. Because of our invisibility, we were able to see more than if we had been large, in a small wildness.

Hungry and chilled, we unpacked our stove in the rain and boiled water. We wanted to get a few more hours of hiking under our belts before building a wood fire and setting up the tent. Within ten minutes, Raven and I sat on drenched logs eating hot soup, watching ground squirrels peeping from behind rocks and watching two deer bedded down in knee-high grass below, themselves like blooms in the minor valley. With wolf-green eyes Raven studied them. One doe faced us, the other stared off in another direction. They were both the red cedar color of summer deer.

For days Raven and I had seen almost no people. This early in the summer, mid-June, snow had not melted from the high country, and we had been unable to reach the Chinese Wall, a stunning formation of layered cliffs that marks a section of the Continental Divide. Thwarted, we had pitched our tent at the site of an old trapper's cabin, where we found a rusted enamel pot, tin cans, and all manner of bottles pitched over the embankment of Rock Creek. A few days we had camped there, writing, painting, and reading around a delicate and constant campfire. Descending to the creek, we'd find a water ouzel dipping its head in and out of the stream.

One dusk a pair of young men cruised past on the trail. I crouched at the edge of a muskeg-like marsh, filtering water, and watched them. Later, the morning we broke camp, we met them, one tall with red hair and the other shorter, with a dark beard. They had settled into the US Forest Service cabin at Rock Creek, the one with log walls scarred from grizzly scrapes, bordered by deep indentations in the duff where bears repeatedly pass, each time deepening their tracks. The young men introduced themselves as biologists, part of an extensive bear survey that would give a more definitive figure on the number of grizzlies left. (In its 2011 report, the US Fish and Wildlife Service reported about 1,500 grizzlies in its Recovery Zone, the Northern Continental Divide Ecosystem of northwestern Montana.)

Now Rock Creek was two days behind us. We had arrived to a bluff flowering with darkthroat shooting star and fireweed above Headquarters Creek, and when we finished our soup, we proceeded down the steep little hill to the stream, a fine place to refill water bottles.

"Let's see how close we can get to the deer," I said.

Raven held a finger into the air. "We're downwind," he said.

The trail was muddy with earth that caked our boot soles and added what felt like a pound of weight to each step. We progressed slowly down the trail because of the mud and because of the deer.

One of the does noticed us. Her ears lifted. We paused, then eased forward a couple of steps. Another set of ears lifted. Raven stopped and I stopped. The first doe rose to her feet. We waited. She stepped backward, the other doe leaped up. We waited. One after the other they ran away from us, up into the bizarre burned country.

"Pretty good," I said.

Creekside, Raven began pumping water for our bottles while I washed pots and spoons. We hadn't brought bowls, to save weight. I was thinking about water, how the air was soddened with it, the ground soggy, the creek swollen. We were collecting it to fill our bodies. So much of life is this one substance, so precious.

The creek was shallow and narrow, not ten feet wide, chortling over its smooth and bejeweled stones. It was lined with willows that had been cropped short by browsing animals. Across the creek the mountain rose into burned and returning trees. I glanced up. Did I glimpse, halfway up the slope, brown movement through a strip of sapling fir?

Our hike had not been without wildlife. On the first day, three mountain goats roamed the alpine rocks of Headquarters Pass. In the days that followed, hoary marmot and coyote, mountain bluebird and nighthawk revealed themselves. Moose and grizzly left their tracks. One sunny day in a deep conifer forest we had come face to face with a bear on the trail. The bear had been fifty feet ahead when we maneuvered a switchback; we could not tell to which of the two bear clans he belonged. We began to back slowly away, eyes averted, and to cajole the bear with sweet talk: "Good bear. We're just passing through. We won't bother you. Nice bear." We continued to tiptoe backward. The bear moved no muscle, but studied us, gleaming in sunlight, very big and very real. Finally he had turned aside and crashed into the woods, and we saw then that he was black.

Now I touched Raven's arm.

"Elk," I said softly. Raven's ballcap kept rivulets of water out of his eyes. He looked where I pointed, his face rugged but serene. Nothing moved.

I shrugged and returned to washing. They were there. I'd seen them.

The next time I glanced up, minutes later, flickers like brown ghosts slid behind a thin scrim of young trees twenty feet farther down. I hunkered beside the creek motionless as I watched. Then a couple of elk stepped into plain sight between trees, first one and then another, and I touched Raven again.

"It *is* elk," I whispered. "I see two. No, three." He sat back on his heels and surveyed the mountainside above us.

"More," he said. "Five."

I caught his eye, raised my eyebrows, and nodded hopefully. Raven went back to pumping and I kept watching the hillside. I could see six elk distinctly now, plus more bronze streaks and flickers. The elk were a few hundred feet away, but they were definitely drifting downward, toward us, and their numbers grew. They kept coming. I realized something.

"They're coming down *here*," I said.

"Where?"

"To the stream."

"Surely they've seen us."

"I don't think so."

Unbelievably, elk poured like molasses behind the scrim of fir. Halfway up the slope the trees breached to make a clearing, and this fanned out in a wide grassy gap. Slowly, cautiously, the first elk stepped out onto the open hillside, onto a broad aperture that led to Headquarters Creek. After a few minutes I whispered to Raven, "How many do you think are there now?" The jangle of creek and murmur of rain muted my words.

Raven raised his eyes and rocked his head in the motion of counting. "Seven."

"I think there are twenty-five." I said it like a confession.

Raven glanced at me. "Twenty-five?"

I couldn't see the elk, only an occasional russet movement, a flowing, but I had been watching long enough to know that animals were pouring down the slope, through the trees. They came and they came, drifting downward one by one. They stepped cautiously into the clearing. Now ten, if I counted correctly, had reached it. They moved slowly, warily, a step at a time.

"They don't see us," I said.

We were crouched by the gushing, ten-foot-wide creek, studded with coppiced willows. We were invisible. I thought about bears and checked behind us. We were safe. There was no hurry. I slowly brushed my teeth and washed my face, and Raven finished with the water-pumping and brushed his teeth. All this time the elk came down and down. Finally we leaned back on our heels and sat in the rain, by the water, below the bank, among the willows. We turned into rocks.

Not one of the elk had seen or smelled or sensed us.

An elk has a shape unlike most ungulates. It has no tail to speak of. Its head and neck are like that of a small horse, its body like a zebra's. It has a long head and a dark neck. Its hide is a richness of browns and reds and tans, moving like prairie grass in wind. Elk hair looks not like watercolors mingled, but like oils. An elk's body has the wet richness of an oil painting.

These elk moved slowly, the older, bigger cows in the lead stopping to raise long dark faces to the breeze: now something didn't smell right, but what, they couldn't tell. The wind was in our favor. One young cow heedlessly rushed forward.

I tallied over and over as elk passed a certain stump or traversed a patch of bare ground—twenty-four, twenty-one, twenty-five.

They were hard to count. Often they moved side by side. Every time I counted, the number changed. I counted twenty-seven.

Apparently the herd was without an alpha bull. Young spikes confronted each other, brandishing short fuzzy antlers, less interested in food than in hierarchy. Typically, one would head another off until the two were facing, then in slow motion, gracefully, the spikes reared at each other, pawing the air like horses but more tenderly, carefully, not ever touching. Much of their activity was this heading-off, then muzzle-to-muzzle contact, followed by a rearing-up and pawing midair.

I could not see Raven's face, only the back of his hat. Turning, deliberate and slow, he whispered, "I bet the ground squirrels are tearing up our packs."

"This is worth it," I mouthed more than whispered. I surveyed behind us again for bears.

"They're very close," he said.

"Yes." Water drowned our words in the small valley.

The cows came down. The calves came. The spikes came. They moved by degrees toward us. We had been squatting now for fifteen minutes and my legs were falling asleep. I shifted imperceptibly, at a glacial pace, onto my behind.

The herd sniffed, came forward, paused. In the background spikes reared and stroked the air intimately, the largest one snuffling and prancing all over the lower slope. The nearest cow elk was thirty feet away. I made no movement, just sat perfectly still.

The boldest cow attained the creek bank, twenty-five feet downstream, and began to strip willows of their leaves. A foot-long branch hung from her mouth as she chewed, then it fell—she bent her long-muscled, sienna-colored neck to pick it up and continued chewing. More cows flowed down and down the hill toward us.

Now one was in the water, the silver creek swirling around her thin, tawny legs.

On the far bankside a cow, browsing willow shoots, let go a stream of urine into the creek. She looked toward us, looked away. Another cow stared at us for a long time and did not panic. Neither of us allowed even our eyeballs to move, but kept our gaze locked forward, thankful for peripheral vision. We were slowly being surrounded by browsing elk. Twenty feet away in the water a spike bull, direct and curious, scrutinized us, but he could not figure out what we were. Over the willow, leaves in his mouth, the bull regarded us. Our raincoats were dark green like young trees. The humps of our bodies were stone.

We were so invisible we were two spirits, crouched by a stream.

The spike bent for more leaves and maneuvered closer. He was fifteen feet away. I wondered then if the elk would finally surround us. How close would they come? What would happen when they discovered we were human? Might we be killed by stampeding 500-pound elk?

The spike was twelve feet from us, in the water, chewing. He could have leaped once and been on top of us. He had forgotten his future as a dominant bull and was feeding eagerly. He lowered his glorious, regal head and snapped off a twig, its leaves tinged red. The muscles beneath his magnificent tweed moved like poetry.

We stared. We stared as long as we could, camouflaged by rain and wind and color and stillness. For five more minutes, seven, eight, nine, then ten, we sat still and stared.

The lead cow, the biggest, foraged past us on the far side of the creek. She was farther away than some of the others, twenty-five feet or so, but downwind now. She breathed in our odor. We saw it happen. The cow started. Somewhere close she smelled rank, unbathed humans, perfumed with woodsmoke and wet wool, or thought she did. Maybe it was polecat. Unsure, she stared and stared, locking Raven and me in her gaze. She chewed, pondering. We were so close we saw beads of water dripping from her chin. We saw the black tips of her big velvet ears. We saw the huge, bottomless, brown orbs of her eyes, yet unafraid, gathering us into them. Her ears were up, her tail flicked gently. She chewed.

Looking back, I remember those moments as hypnosis. I was under the spell of wilderness and under the powerful spell of the elk. If the cow had asked me to go with her, I would have. For this one fleeting interlude, I belonged to her world.

Finally wind brought enough smell that the cow moved backward, head up in an unmistakable position. She pumped her hooves in the acknowledgment that something was amiss. Interpreting her warning and relying completely on her instincts, the other elk immediately stopped pulling leaves and turned as a measured herd and headed back the way they had come. They went orderly, not in a rush, never charging or stampeding. They responded instantly but not chaotically.

Watching them go, I ached. Then, slowly, an animation that I would recognize as gratitude began to fill my body from the top down. The feeling made me dizzy. I thought I would faint.

Even when we finally stood to leave, the elk registered our movements but did not bolt. We climbed the hill slowly to our packs, which neither the ground squirrels nor the grizzlies had bothered, and shouldered them. That night we would pitch our tent in a crevice among black and fallen logs, and by the evening of the seventh day we would be back at the edge of wilderness, and two days after that, not even in Montana.

Raven said later that perhaps we did the elk a disservice, letting them go thinking they could trust humans. Maybe we should have stood and shouted. But I don't think so. I'm glad we waited quietly among them, and saw what we saw.

Montana

Entreat me not to leave thee,
or to return from following after thee.
—*Ruth 1:16*

Coyotes howl at night on the prairie of Wild Horse Island, in western Montana's Flathead Lake. When they woke me, a half-moon had risen and Orion was directly overhead, a sentinel. The coyotes had two camps, one east and one west, and their songs passed back and forth across me in bays, trills, howls, and barks, a night-code. I lay awake a long time not understanding it and also understanding it all too well. A howl is a sound-line connecting those separated by space. I was a long way from my people. I expected to sit up into the prairie and see one of the he-dogs poised, gray in a trickle of moonlight, watching me. I think that happened.

Toward morning I dreamed of my grandmother. She had died, at the age of ninety-three, a few weeks before I'd arrived in Montana. I spent her last fortnight on earth, when she wasn't able to even speak, beside her bed. In the dream I was haying a field on her land. I was on the old Massey Ferguson, and the sun was close and brilliant but not oppressive as it usually is in late summer in the

South. My arms were bare and I was practically flying over the field. I knew that although my grandmother was dead, she was watching from a line of water oaks at the edge of the pasture, yet her body was the hayfield itself. I dreamed this so wholly that when I woke, I thought I was in my grandmother's grassy arms.

Newly arrived to the West, I'd left the swamps and pine flatwoods of south Georgia, where I was born and where I was encircled by layers of familiarity, and I had enrolled in graduate school 2,500 miles away. This was long before I met my husband and married.

"You won't be back," my friends said. "You'll get out there and fall in love with the West and with one of those cowboys."

"I'll be back," I replied.

Now I was with a professor and fifteen colleagues, set to spend a fall weekend on an undeveloped island, a couple thousand acres of wildness. Although public camping wasn't allowed, Hank Harrington, my professor, was one of a handful of private landowners, and we'd be camping on his property.

Our assignment was to choose one animal species and follow it for two days.

Hank led us to a Palouse prairie at the heart of the island where two hulls of settler cabins persisted. "Wild Horse Island is shaped like a fish, north to south," he said. "You're standing in the narrowest spot." To the east rose a fan of low island mountains. To the west, in the distance, was water, then the mainland and the Salish Mountains beyond.

"If you get lost," Hank said, "follow the lakeshore back. If you get hurt, we'll find you. Okay, then. See you."

I headed up the nearest hill, which overlooked a bay I'd seen on a map, Skeego. As I shuffled through the flowering prairie grasses I wondered how *Palouse* was spelled and what it meant.

The ecosystem resembled the savannas of the southeastern United States, except the grasses were foreign. Back home, I could name them: wiregrass, bluestem, toothache grass. Here I was lost. Not knowing their names was like having music stolen. The world was flatter, duller.

The slope down to Skeego blurred blue-green with Ponderosa pines, but what I thought about was longleaf, the pine of my home, which had historically dominated southern uplands but was rapidly being logged.

There was also the matter of the island itself. Just before I left Georgia, I made one last trip to Cumberland, a wilderness barrier island off the coast, a national seashore. Sitting on a high bluff above tidal Christmas Creek, I'd seen so many wading birds, mostly white ibis and snowy egrets, that water lotus seemed to bloom across the mud flats. A mother raccoon with three young passed through my camp, followed by a flock of wild turkeys. Wild horses allowed me to pass within twenty feet of them, and at night I could hear coyotes singing along Crooked River. The West was renowned for grandeur, but the South coursed through my blood.

My six-year-old son, now back in Missoula with friends, felt the same. A few days before he had said, "Know what I miss?" and I expected him to remind me of some toy we'd packed away or how he used to steer the truck for Grandpa.

"Cypress trees," he said.

I ran into pygmy nuthatches first thing. I had been giving myself a crash course on the birds of Montana, and I recognized this one. A Ponderosa pine was alive with nuthatches, as if it wore dozens of dangling earrings. The birds were feeding mostly from cones but also from beneath the flaking bark, tossing bits away with little rips of their heads. Their tapping kept the tree chattering.

I thought about the assignment at hand. Did I want to follow pygmy nuthatches for two days? Did they have something to teach me?

Maybe.

I headed eastward along an animal trail toward the north shore. A belted kingfisher shrieked away. Should I follow the kingfisher? Two dark-eyed juncos hid in a cedar loaded with light-blue nuts. The trail, bordered with purple asters, started up a rise, and after a couple of hours I came to a damp woodland cliff. The cliff was threaded with fine ungulate trails, and harebells bloomed along one steep, cool face. I saw from the map that the terrain got steeper and rockier. I headed upward, toward the highest peaks, watching for bighorn sheep, bear, mule deer, mountain lion. These large animals would be harder, even impossible, to find and follow. Squirrels bounded and flashed, leaving stump-faces littered with pine cone scales like piles of artichoke leaves. My breath said, "Flatlander, flatlander, flatlander."

In this manner I explored the island.

Late in the afternoon I walked out of the hills onto a high, grassy meadow. A bird in a pine a few feet away spooked and circled over the now-visible basin of Flathead Lake. Onyx lines of tears dripped from its eyes, and its tail flashed russet in the lowering sun. It was a kestrel.

Kestrel, I thought. Back home we had these. I felt like a half-hearted traveler in an inconceivable country who seizes one familiar word. *Kestrel*. If I could erase the chain of mountains that swept away and replace it with fields of broken cotton stalks, with the kestrel perched on barbwire, I'd be home.

Back in Georgia, my good friend Milton, naturalist and farmer, recently had seen the first kestrel of the fall. He told me in a letter.

Every winter both of us had awaited their return south. Driving together through the bleak, frostbitten farmland, we'd make a game of counting them. Sometimes there'd be a dozen sitting on power lines between Osierfield, where Milton lived, and Baxley, my hometown an hour away.

Now a second kestrel, one of the most striking birds on the continent, landed.

I sat down and pulled off my boots to pick some burrs (cheatgrass, I later learned) from my socks. I was on a grassy mountain flank strewn with gray boulders made of what seemed like petrified filo dough. The rocks were mottled with at least four colors of lichens: sea-green, mustard, chestnut-brown, and the green of old dollar bills. The two kestrels converged on a forty-foot pine snag. Every two or three minutes one of them swooped toward the ground, then returned to perch. They appeared to be hunting grasshoppers. One kestrel touched the ground, paused, then lifted and banked wildly a yard above the ground. When she flew back to the tree, she landed on a different perch and ate, bobbing her head, picking at the food in her talon in a sedate and elegant way. A third kestrel arrived. I started to scribble field notes.

All afternoon the day after Grandmama died, people gathered in a fluctuating circle under a water oak in her swept yard, waiting and listening to stories. Folks came and went, bringing food—chicken and dumplings, broccoli salad, a pan of rolls, pound cake. No one cried. We sat in the presence of death without mentioning it.

The kestrels fed farther and farther up the mountain until they were out of sight. Barefoot, I followed them, thinking to get above them before I crept across a ridgeline. A sharp stone punctured the sole of my foot. For all my calculation, I rose too close to the kestrels. One bird broke from a pine and flew out over a shady canyon.

Another disappeared. A band-winged grasshopper sputtered away, red wings spiraling. The mountain felt soft and warm to my tired feet, and I thought about how a person gets to know a place. It is a reciprocal process of incorporation, of adding one life to another. It is an exchange.

First you bathe there. My son and I, arriving cross-country to Missoula, had stopped east of town to swim in the cold Clark Fork River, and this morning I'd plunged into the glacial lake when we first arrived.

Then you go barefoot, naked if you can.

Then you eat and drink from a place. You sleep with it.

You watch and listen. You study. You learn. You listen.

Then you stay.

Two kestrels lifted from a grassy meadow below, a tiny bowl divided down the middle by an ungulate trail, similar to the line pregnant women get on their bellies, and then the birds floated toward a clump of Ponderosa pines growing in the drainage. When they flew they called *killy, killy, killy,* before going silent. Now I was higher than kestrels, breathing in the fragrance of butterscotch and sandalwood. I could see the backs of their airborne bodies.

The sun began to set. Where it struck Flathead, a gaseous ball of white-hot fire seemed to boil the water. The mountain ridge on the mainland, worn and blue, fenced the sun in the sky. Mule deer warily lifted their heads from late grazing, and ten honking Canada geese, winging home, sealed shut the day.

Late afternoon I ironed a nice blouse, used a little of Grandma's makeup, and rode to the funeral home with my first cousin Jimmy for the viewing. Grandma's casket had been wheeled to the front of the chapel. Shining silver, it was surrounded by wreaths of carnations and huge pots of white lilies. She had loved flowers. The casket was open, and the top of Grandmama's head, lying against pink satin, was visible. It looked as if any moment she would raise herself and ask what on earth she was doing in such a ridiculous box. "Help me out of here," she'd say, and I would. Her eyeglasses rested on her lifeless face. What need did a spirit have of glasses?

Only a few people, all immediate family, were in the chapel. They sat contemplatively on velvet cushioned benches. I went and stood beside Grandma, looking in, and Aunt Coot joined me. We had often stood together beside Grandma's bed. My aunt reached down and touched her mother, smoothed her hair, fingered her elfin ear. The ear was cold, frozen in place. The undertaker had fixed Grandma's hair special and she would've liked it. I touched it. It felt as it always had, rough and wiry. Her hands were cold. A few days before, I'd removed fingernail polish that one of the nurses had painted on. My grandmother had not been a nail-polish kind of woman.

We sat and sat. People started to come—people from Spring Branch community who'd known Grandma all their lives. Neighbors. Nieces and nephews. Distant kin. I tried to meet them all, following genealogical lines, fitting people into a framework of history and place that included me. "You have to be Lee Ada's daughter," someone said. "You look just like her."

During the funeral the next day I sat beside Uncle Percy, the son who didn't leave the homeplace. Most of the time he twiddled his thumbs round and round, motion without purpose, but once I looked over to see a tiny spring flowing from his eye. On the other side, Aunt Fonida's body rocked and shook with silent tears. Uncle James, the eldest, a tall and stately Baptist minister, recited Grandmama's favorite Bible verse, chapter 1, verse 16 of the Book of Ruth: "And Ruth said, Entreat me not to leave thee, or to return from following after thee: for whither thou goest, I will go, and where thou lodgest, I will lodge: thy people shall be my people, and thy God my God."

*I rode with Aunt Fonida in a slow burial procession, following
a glossy black hearse from Spring Branch Church along a clay road
that turned toward Carter Cemetery. Behind us a long queue of cars
snaked past the houses of Grandma's neighbors, farms she'd passed
all her life, places she'd stopped to visit. She took her last journey
of this world, back to a clay hill where she would be buried beside
Granddaddy, among all the dead of that country.*

That evening in Hank's cabin on Wild Horse Island, we grad students drank a lot of wine and told a lot of stories, mostly urban myths. When I dragged my sleeping gear into a dark field, the sky was a hand-painted bowl, silver on black. The grass was dry and stringy. To the east an improbable amorphous glow crowned the tallest mountain. Was that the Pleiades? Could it look that huge? The Big Dipper hung on its black wall, above the Little Dipper. The Milky Way was a sash tied across the revolving sky; it was so thick with stars, I am not kidding, that it looked like cottage cheese. As I stared up, an owl's dark shape winged noiselessly overhead. Not long after I went to sleep, the hooves of a galloping wild horse, thudding against the ground, woke me. The second time, I woke to coyotes, the third time to a dream.

*My grandmother had loved birds. For years she was the only person I
knew who kept bird feeders, homemade ones Uncle Percy fashioned,
just outside her kitchen door. "Won't you look at this purty little yellow
bird," I'd hear her say. Or red bird. Or blue. I used to spend Saturdays
with her, and those days at her farm would be long and slow. I lived on
a junkyard at the edge of town, where my life was very different from
my grandmother's. One noon, as we cleared up after dinner, she came
through the screen door from throwing scraps to the dog. "Listen at
that," she said in her sweet and royal way, not just smiling but almost
smothering a laugh, such was her glee. "That big old woodpecker is*

just a'laughing." A pileated clung to the old-growth longleaf in the
backyard. I knew no one, apart from my grandmother, who would
even notice, let alone take pleasure, in a bird call.

Evenings, we'd sit and rock on Grandmama's piazza, trying to
stay cool. The edge of her porch, outside the screen, was lined with
impatiens and geraniums. There'd be a whir and a ruby-throated
hummingbird no bigger than a Fig Newton would flash among the
faces of blooms. "Bless gracious," she'd croon. "You're a fine little fellow."
Then it would purr away.

At dawn I set out toward the canyon where I'd last seen the kestrels,
soon passing through an abandoned orchard. No more than fifteen
trees grew in the orchard, mostly apple but some pear, and all the
trees were thick with fruit. Beneath the trees the grass was like a rug,
beaten down. On the ground not an apple could be found, although
at the very moment I entered the grove, a wild apple hit the ground.
Small as a baby's fist, it seemed both a gift, sour and delicious, and
an invitation, so I shook the tree and gathered a handful of fruits,
gnawing at them as I walked.

A coyote bounded toward the drainage.

I came out onto the prairie. Five mule deer watched from a swale,
probably attracted by the thudding of fruit. Their antlers were silhouetted against a pink promise of sun. Meadowlarks, the western
variety, birds truly gifted with voice, sang. Then, in the middle of the
meadowlarks' performance, a coyote howled. It was close. I knelt in
the grasses that were fat and silent with dew. I could see the coyote
through the trees, poised on the next hill. It began to sing, long and
mournful howls like train whistles past lonely boardinghouses and
run-down shanties. Below, in the drainage, another coyote, then
another, barked in reply. A yelping commenced, punctuated by the
long, plaintive yowls of the coyote in view. On hands and knees I

inched closer to her, staying behind sapling pines, thinking about the coyotes but also about fire and how it maintains a prairie, so that it stays crawlable.

I knew that my presence—early out, the sun no more than golden streaks brushed above dim, dark-green mountains—signified something to the coyotes. I knew that their conversation told of my being as much as it told of their own, as a hollerer in the great Okefenokee Swamp back home sang not only to locate himself to his neighbors, but them to him, and even him to himself. I was connected to the coyote through lengths of song-rope, and to hear myself in animal song made me feel humble and blessed. I knew that in one brief visit, in vanishing grasslines my feet laid against the earth and in the unrecorded songs of coyotes, I had become a tiny, faint part of the island's legacy, one of its brushes by wind. And neither could I go free of it. It had also brushed me.

Not long afterward a kestrel landed at eye level. It simply sat, waiting for the dew to evaporate and for the sun to warm grasshoppers into motion. I circled toward a rocky cliff, above the waiting kestrel, on the western side of the deep canyon where I had observed the birds the evening before, and I soon saw that it was a good place. Two kestrels perched in a snag intertwined with a shorter pine, the dead tree cradled in the live tree's arms. The trees were rooted where a talus slope leveled into a run of rocks, so that the pine's upper branches brushed the craggy cliff.

When the sun topped the mountain I had to shift positions, crawling up the slope so as not to be blinded. Four or five bighorn ewes, bedded down fifty feet above, watched without panic. From the canyon bottom I heard a clatter and looked down to see a hugely antlered mule deer. Though many hundred feet below, it struck into a run when it spotted me, its hooves raising a mantle of rock fragments. Another, following a few minutes behind, repeated the maneuver. Next, two kestrels flew out from rocks and performed an aerial stunt, flying in unison, making a twirling W, then separating. I watched through binoculars—feeling slightly sick, since the glasses had the effect of suspending me in the abyss—and tightened my feet against the rock overlook.

All morning I watched kestrels flashing copper tails, arrested in flight, the sun washing their slate-gray backs to midnight blue. The fresh fall sunlight laid yellow-green ribbons down the backs of the ridges and dusted the meadows gold. A kestrel landed and clung to a mullein stem. The dried stalks stuck up like birthday candles, and here and there grew a new rosette of hairy leaves, doomed to be killed by coming frost. At one point I counted seven kestrels feeding in the Palouse prairies of Wild Horse Island.

After Grandmama's funeral, after everyone had eaten the last helpings of chicken-and-rice and pineapple cake back at the house, after they fled to their new cars and left all that history no longer relevant, only Mama, Daddy, and Uncle Percy remained.

I watched until the kestrels spread across the morning. The ewes vanished. The sun pasted itself to the top of the sky.

Before long most of the kestrels would be gone south, out of winter. I would not be leaving. I stumbled, then folded into a niche of boulder that overlooked the magnificent chartreuse relief of Wild Horse and wept bitterly. My grandmother would not be there when the kestrels returned to their southern home.

The morning had been a movement of such beauty that at times I had gasped aloud. I had come from wayward sleep, sat in meditation with bighorn sheep stretched behind my shoulders and rock wrens at my feet, and watched a suite of falcons in panoply. Gradually this place—this stretch of the Rockies, with its enthralling bighorn and bawling ravens, its chilly days steeped in the amber vestiges of summer—sidled its way into my heart.

After a while I stood and tied my sweater around my waist. Rent in two directions, full of longing for both the past and the future, the unfolding mountains and the backdrop of home, I made my way to the cabin. That's when I knew I would fall in love with the West and that I would be devoted to it.

But I could not stay.

POSTSCRIPT: In January 2008, eleven years after I left Missoula, Hank, along with his wife Nancy, drowned in Flathead Lake while attempting to navigate a canoe crossing from Wild Horse Island to the mainland. The boat capsized in high winds, dumping the Harringtons into frigid water, and the two were unable to return to their craft. I am grateful for Hank's life, for what he taught me, and especially for him taking me under his wing when I went west to study.

Opening the Big W

From the wind-tortured summit of Mt. Henry, from its creaking lookout, Canada is dark blue and misty, eight miles away. The lookie points it out. He's a young man named Leif Ericksen whose eyes are so blue that they look like the sky itself, cornflower blue against the royal blue of Canada, shining through augured sockets. Today, over in Canada, it's raining.

In the United States, in the Yaak Valley of Montana, it's plenty sunny, and the sun bores into the lookie's eyes, into the Purcell Mountains, into places it shouldn't reach at all. The sun pierces into gigantic lime-colored holes in the evergreen forest.

Clear-cuts.

The Forest Service calls them "openings."

There are so many of them. Looking due east and using peripheral vision, I count fifty-two. Sometimes they have a smattering of trees—more recent clear-cuts—but many of them are wide grass patches, White House lawns in the Montana woods. They are like ghettos. The strange thing is, this is public land, most of it part of the Kootenai National Forest. The people's land is being logged.

"How big is this littlest one?" I ask Leif, who leans over his ramshackle widow's walk. I could tromp down to this cut in twenty minutes.

"About forty acres," he guesses.

One of the clear-cuts is six times that big, an entire mountain flank reduced to stumps and grasses. In fact, on these closer slopes, the amount of cleared land almost equals forest, and threaded through it all, like tapeworms, are skinny white logging roads.

Don't mistake what I am saying. On the darker-green slopes beyond, there are clear-cuts. Behind, to the west, dozens more are neon green and blinding in the sun. Sometimes only narrow swaths of trees separate them, corridors that look to be the width of rope footbridges and about as secure. *In every direction there are openings.*

A man named Rick Bass lives in the Yaak Valley. He moves among the trees quick and wary, out of sight in a wink. Many times you don't see him at all, but you know he's out there, in the woods. Like you know the mountain lions are there. And grizzlies.

Rick wrestles with the openness, although *wrestle* is too mild a word. The cutting rides him hard, like a disease, because he lives tight in those trees, lodgepoles and Doug firs and larch crowding up around his house so close that you hardly notice the house at all.

His life work has been not only his writing—his fiction—but his dedication to protecting the last roadless areas in this wildest of places. That's how I know him.

Rick's a broad-chested, tough man, not tall, with thin brown hair and intelligent eyes. His face is so open that he has to work to control the emotions that traverse it. He came to the Yaak as a young man, and every day he wakes up more twined in the forest, his body woven into brown-gray bark and gray-brown branches. He tells me that the first time he saw the valley, from a high point

on the Yaak Road, coming from Canada, something snagged his
heart, like an oversize fishhook. He showed me the spot, rounding
a curve, slowing, the Yaak laid out like Sleepy Hollow.

"This is it," he said.

People call it passion, but it's more than that. A person cannot
tell where Rick's body stops and the valley begins. The screen separating him from wildness is translucent, it is so thin. It seemed as
if before my eyes Rick Bass might break free of domesticity and
shape-shift to wolverine. I couldn't keep my eyes off him, wanting
to witness the transformation.

"I get scared looking at the clear-cuts," he said. "I'm afraid they'll
make me sick. I mean really, physically, some disease of bitterness."

At Rick's suggestion, I've climbed Mt. Henry to look around.

"I see plenty of clear-cuts, but where are the roadless areas?" I
ask Leif, who has begun to prepare his supper, bag of oregano in
hand. A man of minimalist glossary, he gestures south.

"Basin Creek," he says. He thumbs generally behind him. "Rod-
erick. Saddle Mountain. Big Creek."

"All we see is the Yaak Valley?" I ask.

"Yes," he says, and that's it.

Some of the hardiest souls in the country live in the Yaak, in a
handful of villages—Yaak, Troy, Libby. Here snow flies from Sep-
tember to May, although one Fourth of July when I was visiting the
town of Yaak, the northwesternmost in Montana, snow was falling.
Even now, late July, a store of unmelted snow laces Mt. Henry's
summit. I think of Yaak as a Rip Van Winkle kind of place, asleep a
long time, buried. But magical.

Some of the people who live here belong to the Yaak Valley Forest Council, Rick included. The culture here is logging, and nobody
wants to shun that salty identity, but some folks are afraid of too

many trees being cut too fast, of losing the wolverine like they lost the woodland caribou, of losing the whole damn place.

All total, 150,000 roadless, unlogged acres are left in the 2-million-acre Kootenai National Forest, not in one big chunk but in six or seven pieces. The Forest Council wants to keep those areas roadless and unlogged, turned into capital-*W* wilderness, meaning protected by statute from human interference, as opposed to little-*w* wilderness, which they are now, unspecific and vulnerable.

I look wistfully into the blue-green mist of Canada, which faces the same dilemmas: development, logging, and growth. Versus wildness, peace, and beauty. Or versus grizzlies, trout, and loons. I've flown over the behemoth and immensely distressing clear-cuts of Vancouver.

Once, at a campsite in the upper Yaak, a woman with long gray hair and catlike eyes roared up in a pickup nearly as old as she. She cut the motor and stalked over to where a friend and I sat in grass beside a little creek.

"A bear lives up here," she glared accusingly. "She's been here for years. You have to be careful with your food, lock it up at night. If the bear becomes a nuisance, they'll take her away. And there's nowhere else for her to go."

"We're being careful," we said truthfully, but the woman was severe.

"If you leave food out, that's not the bear's fault," she said. "This is the end of the road for her. There's nowhere else to go."

The Yaak Valley is the edge of the country, pushed up against Canada, and if it gets laid open, austere and soulless, then we can give up on wildness in this nation of fools.

Some people think the Yaak's already lost. The head of one environmental organization said to me, "Have you seen it? It's been cut to death."

I'm looking at it now. I'd like to lay my own place, the coastal plains of south Georgia, up next to the Yaak. We have precious little public land, and our forests have been devastated. In terms of longleaf pine, which covered the upland South, we got down to 1 percent; the rest was turned into housing developments, parking lots, trailer parks, big-box stores, and pine plantations. Some people think southern pines naturally grow in rows. Ninety-three million acres of longleaf once covered the uplands of the Southeast, and 99 percent of it is gone. Only a couple thousand acres of virgin longleaf are left, anywhere. If I could lay that survivor landscape across the northwest corner of Montana, you'd see why I'm not writing the Yaak off.

In 1995, then-Congressman Pat Williams introduced a bill to designate 42,000 acres within the Yaak River drainage as Wilderness. The word *Wilderness* scares a lot of people. They think it means they'll have to Butt Out and Keep Off. To them, Wilderness steals jobs from their kids. The bill was defeated.

"What you've got up there is a working landscape," Hal Salwasser, a tall, impassive man who was the regional director at the time, told me. I had gotten on a first-name basis with him after running into him at the Missoula airport and striking up a conversation. At this particular moment we were in his Missoula office, sitting at a lustrous boardroom table. He had returned from a visit to the Kootenai.

"Why do you need Wilderness in Yaak?" he asked rhetorically. "The Yaak's a perfect example of a landscape that works—you've got logging, you've got grizzly, you've got inland redband trout. You've got all the pieces."

"But will we have them in five years, at the rate we're logging the place?" I asked. "Will we have them in ten?"

"Definitely."

"*Somehow* the wild creatures are disappearing," I said.

Take grizzly bears. When he arrived in the Yaak, Wayne Kas-worm, bear biologist with the Fish and Wildlife Service, found the grizzly situation "so tough that unless we did something we ran a risk of losing them." About thirty-five bears roamed the Yaak, with less than a hundred in the whole Recovery Zone, as it's called, which covers the Yaak Valley and the nearby Cabinet Mountains (where there *is* designated wilderness).

Historically, up to 100,000 grizzly bears ranged west of the Mississippi River. By the early 1970s they were down to 1,000, ten times a decimation, or, said another way, decimated ten times. By researching mortality records Kasworm found that from 1950 to 1990 at least sixty-five bears were killed in the Yaak.

In 1975 grizzlies were listed as endangered and hunting ceased. Although four females from British Columbia were introduced to the Yaak in 1990, by the middle of that decade Kasworm was "hard pressed to say their population has changed dramatically." As of 2016 he estimated fifty bears, and the figure is now in the low forties.

Bears shy away from areas of high human use, such as clear-cuts and roads. A typical female's home range, as it's called, is between 75 and 100 square miles, a male's between 300 and 500. What happens when animals get boxed in?

People sometimes confuse Wilderness with Government Interven-tion. They sometimes are led to believe that when Wilderness is granted, free will is taken away. In those years the Forest Service was closing hundreds of miles of roads, some for grizzly recovery, some for elk feeding, some for watershed protection. People didn't want to lose their privileges on the land: picking berries, hunting, fishing, and gathering firewood. In response, residents destroyed gates. Outside the Silver Spur in Troy I saw a bumper sticker that said *If Roads Are Closed in Heaven, Then I'm Not Going.*

That people could still walk, ride horseback, and bike the roads, or that thousands of miles of roads *were* still open in the Yaak, and millions of miles in the United States, and more opening all the time, and widening, didn't matter.

The constituency for Wilderness, however, was growing. More and more people were seeing that designated wilderness is a grand idea for preserving landscapes, ecosystems, biota, the atmosphere, cultures, traditions, ways of life, escapes. They were understanding that wildlands could be protected forever even while logging continued, if that logging was ecological and sustainable.

One year Rick arranged for the Orion Society, an organization that advocated for nature via literature, that published *Orion* magazine, to bring the Forgotten Language Tour—nature writers reading their work—to the Yaak. This would be the first tour to take place in small-town theaters and community centers rather than universities. Terry Tempest Williams (author of *Refuge*) would come from Utah; Richard Nelson (*The Island Within*) from Alaska; butterfly expert Robert Michael Pyle (*The Audubon Society Field Guide to North American Butterflies*) from Washington. Rick invited me.

Four nights in a row we read—in a walk-in theater in Troy, in a church in Libby, in a community center in Yaak, in a lodge near Bull River. At a visit to a two-room log school in Yaak, Richard Nelson asked a handful of children: "Who saw a wild animal today?" Every hand went up: raven, deer, squirrel. The children led me to a little stream that runs behind the school and showed me caddis flies in stony wrappers. One day we hiked with high school students in the forest behind a little rail-fenced cemetery in the upper Yaak. Rick found a morel growing out of a grave in a corner of the cemetery. *He lies in the woods he loved*, the headstone said.

A storm happened while we were touring in the Yaak. It was a storm so powerful and erratic, so unexpected and dreamlike, that for weeks people would tell each other storm stories.

We had journeyed to see old-growth cedars up Seventeen Mile Creek, in multiple vehicles, but Rick, our guide, hadn't yet arrived. We milled around waiting for him on a narrow access road, in a stand of thin lodgepole pine, talking and enjoying the woods. We noticed that the sky to the southwest began to darken. Then lightning began to spill out across the mountains. Thunder clapped.

About the time Rick rushed up a sudden wind coiled through the trees, taking hold and building in intensity. In a matter of minutes fifty-foot trees were bent double, whipping each other with their crowns, shrieking in the high wind. "I marveled that the trees were so limber," someone said to me later. "I thought that was normal." Drops of cold, high-country rain flew through the pines and struck us, hard. It hurt.

When the first tree cracked, we scrambled for cover. Nobody ignores that sound. We rushed for the vehicles.

Now trees were thrashing, falling all around us, and rain was twisting from the sky in sheets. No sooner was our caravan lined on the access road, moving out, than a tree toppled in front of the lead car. By this time hail big as sugar peas tumbled out of the freak storm cloud, making a tremendous din against my truck roof. It was the kind of hail that strips skin off you. I could do nothing except huddle in the vehicle, hoping I was safe from falling trees, hoping the storm would soon be done.

The microburst subsided as fast as it had come on. I gingerly opened my door and stepped out into a layer of ice pebbles on the ground. Trees were down in front of us, behind us, between us. But nobody was hurt, not a car touched. Most amazing was that something told the driver of the rear vehicle to back up, and he did, seconds before a tree crashed down where his van had been.

"Something told me to back up," he kept saying.

I'll never forget Rick's reading at Bull River that evening. The storm had knocked out power in the valley, and he was reading by candlelight, a short story called "Swamp Boy." Tiny shadows from the candle flicked across his face. The story was about a boy who loved the woods, who could not get enough of them. The candle shapeshifted into a fire we sat around, listening to a storyteller who knew how to traverse one world in order to enter another and bring it back in description and metaphor. All around lay illimitable darkness. Rick's voice broke. He was Swamp Boy. I was too. I was moved to tears by the events of the day, by the beauty of the evening, and by the power of the story. Tears ran down my face. Swamp Boy was alive.

Rick never stops, I don't think. From morning to midnight he's working for the Yaak, not just to preserve the landscape but to protect the animals and the people. He writes stories, he writes letters, he goes to meetings, he guides photographers. Even when he's tired he won't stop. Last time I was up there he was coughing, coughing, and not slowing down at all.

Was Rick tapped to do this work? Was he called to speak up for wildness, to put his body on the line for it? Was he marked for it from birth, knighted? Was he born so wild that the destruction of wildness also destroys him?

"There is so little uncut country left in the Yaak," he says, "we need places where there are, for once, no stumps—no matter whether

horse-logged, helicopter-logged, tractor-logged, or cable-logged—
just a place for our minds and bodies, our eyes, to rest." Until then,
the weary warrior slips through Yaak's sickening openings, waiting
for a simple designation that will change all our history forever.

One Meal

I thought we'd make a hike, as Rick called it, and climb some rocky sylvan trail sweetened with Indian paintbrush, sensate with the promise of summer. That would have made more sense.

When I pulled up at his timber-frame house, Rick was unloading grocery bags from a beat-up truck. He disappeared into the house and came out a minute later with his arms free.

"Well, hey!" he said, with a hug. "I thought you got lost."

"Hey back. Just going slow, trying not to hit a deer." Rick's little girls poured out then, inches taller and degrees cuter.

"Look at you two," I said, hugging them, two giddy little kids whose eyes gleamed. "Where's your mom?"

"She's gone with Grandma," Mary Katherine said.

Rick's mother-in-law was visiting, and his then-wife Elizabeth had taken her sightseeing. "They'll be home by supper," he said.

Although summer had almost arrived, the day was cool and overcast, drizzling on and off. The weather hadn't made up its mind about what it wanted to do. At any minute either rain would pour down or the sun would emerge brilliant and marvelous.

"Let us help you," I said to Rick. I was thinking if we got his chores done, we could leave sooner for an exciting trailhead.

There were a lot of grocery bags. We ferried them into the house. My conversation with Rick had already started where it had left off when I last saw him, writing and writers, activism and activists.

"Well, we have all day," I said. "Where should we go?"

"I thought, since it's wet out, maybe y'all would like to cook," Rick said. (He's originally from Texas, which explains the *y'all*.)

It sounded like a joke. Maybe we would like to cook?

The problem with cooking was that Rick was not a cook—he never mentioned cooking in his letters—and neither was I. We were writers. The other problem was that there was a whole wild world outside, full of fairy slipper orchids and mountain goats. And the biggest problem: I preferred the wild world.

"I think it's going to clear up," I said, looking out the window at hemlocks and firs beyond the stone wall.

"I found a recipe for a chile-crusted grilled turkey," Rick said.

Jesus Christ, this didn't sound like Rick Bass, the man who wrote stories about crossing a mountain with half a loaf of stale bread and a jar of crystallized honey in his rucksack.

"Are you cooking these days, Rick?"

"A little," he hedged. "I decided to buy game hens—a turkey's too big." The conversation was moving faster than my molasses brain could handle. Rick hoisted a cookbook already open to a page and showed it to me, as if a book would lend credence to this outlandish scheme he'd hatched. *Tom Douglas' Seattle Kitchen* was the book. The guy had won all kinds of awards.

Rick was sounding more like Rachael Ray than Ed Abbey.

"Girls," I asked. "Is your dad cooking these days?"

"No," they chimed, loving the intrigue.

"What if your mom's away?"

"Macaroni and cheese!" Mary Katherine said.

"Look, Rick," I said. "This weather could turn on a dime. I've got apples and a bag of chips. Let's go take a hike." *Take* instead of *make*.

"I really want to try this recipe," he said. Maybe there was a side to my friend I didn't know about. Maybe he was more than wolves, winter, and wildness.

"Really?"

He pushed the book toward me.

Chile-Crusted Grilled Turkey. The recipe had three parts to it, one for a stuffing and another for something called a rub, which must be what the *chile-crusted* referred to, then the part where you actually cooked the turkey or, in our case, game hens. Any recipe with over ten ingredients was too complicated. My success rate for scorching food is at least 50 percent.

I looked out the window. The fir trees collected baskets of silver droplets, nothing more. Along Vinal Lake lupines would be in bloom. Loons would be calling.

I met Rick while I was in grad school in Montana. At the time I was a single mother studying nature writing, and Rick was scheduled to do a reading at Freddy's Feed and Read, as it was called, a little bookstore in Missoula that nabbed good writers because of its proximity to campus and because it cared about literature. My professor, Hank, was set to introduce Rick at the reading, but my luck runs toward nirvana.

"You around on Sunday?" Hank asked me midweek.

"I plan to be," I said.

"You want to introduce Rick Bass?"

"Are you serious?"

Hank laughed. "I am," he said.

"Yes, yes, I do, yes."

Introduce Rick Bass? I'd give my wooden kayak paddle to do it. And my iron skillet.

For the next few days I felt like a writer among writers. A lit match had been tossed into dry tinder inside me. Back in Georgia, I'd read *Wild to the Heart* and everything else I could find with Rick's byline on it, and now not only was I going to get to hear him read in person, I was going to introduce him!

Rick seemed to like what I said. He was nice to me. He gave me his address, and after he went back home to the Yaak and I went back to class, I wrote him a letter, and he wrote me back. I wrote another. He responded. That went on for a while. Some of Rick's letters were just a line or two, sometimes on a postcard, because he had hundreds of people like me in his life, and my letters were long treatises on writing as activism and whether it was better to be a hunter-gatherer or an agrarian. I didn't have anybody else I could talk to about these things, at least not somebody like Rick, who was thinking about them in a way I very much admired and wished to emulate.

Because of these written soliloquies, Rick invited me up to do a reading in the Yaak. Honestly, I think he did what anyone would do for someone who was burning with a passion similar to their own. He encouraged and helped me.

Then grad school was over and I went back to south Georgia, but not before I sent more letters and made a few more hikes. A year passed and we kept up a correspondence, his letters often quick, but full of beauty and pathos and wisdom, and mine embarrassingly long and about my dirt road getting paved, litter in the ditches, nesting Carolina wrens.

Now I was back in Montana for a visit. I'd written Rick and in the crazy beaver dam of his schedule he had one day that I could come up to the Yaak and we could make a hike. *Make.*

Since Rick arrived in the Yaak Valley a few decades ago, he witnessed the government-subsidized clear-cutting of the national forest that surrounds his home, a country of elk and grizzly, a stunning tableau of long, deep, brave winters and short, glorious summers. He identified the last roadless areas and worked to get them protected. He'd written a lot of books about his home, and I'd read all of them.

I was an environmental activist too, but the difference was that Rick worked a hundred times harder. He worked tirelessly, ceaselessly. He got up at two, three, four o'clock in the morning and went to his office to write letters. I liked to sleep.

Friendship for me has not been a glory-slide through an amusement park. I never learned to have friends as a kid, not because I was shy or unfriendly but because my puritanical upbringing forbade it. I could be friends with God, but pretty much nobody else was worthy. I never had a sleepover, never had a play-date, never had a date. When I went away to college, I was socially retarded. I had to start at square one.

I discovered quickly that I liked people. A lot. I usually didn't find them any more amoral than I, or inadequate to receive my attention. In fact, I found myself immeasurably and incurably curious about people—their lives, how they lived, what they thought about. The most shocking thing I learned was all the ways I was similar to them. And how much I enjoyed doing stuff with them.

Because friendship was an anathema and because so much goodness had been required of me as a kid, I judged myself mercilessly. As a friend, did I measure up? As a writer, as a naturalist, as an activist? Rick had written so many books and I'd written none. He'd started a nonprofit, I'd started a garden.

Now, standing in Elizabeth's kitchen, I looked at Rick's girls, who were following the proceedings with bright eyes. Would we explore or would we cook? Mary Katherine, sweet child, was reading the recipe over my shoulder. I turned and made a face at her. She raised her blond eyebrows slightly and suddenly filled the kitchen with peals of laughter. Her father, the man whose specialty was peanut butter sandwiches, intended to cook?

Rick had collected a raft of ingredients on the island—frilly green plumes and golden globes and bottles of red powder, microscopic brown seeds and oranges.

"You're not doing the math," Mary Katherine said to me, ringing her bells of laughter again. By that pet phrase she meant that I had misjudged her father. We'd be inside. The younger girl, Lowry, doll-like in her beauty, understood fun.

"Okay," I said, my eyes on the girls but speaking to Rick. "We'll help you. How do you need us?"

I've realized a few things: really good friends are hard to find. These are usually people with whom you have a lot in common, who share your visions. Most friends don't last forever, they last for a time, glued to you by geography or job or interest, and the glue dries and cracks, the friendship falls away, like a satellite, drifting off through the universe.

I lost a few friends because somebody moved. I lost a few to ethics. I lost a couple to romance and some to my own stubbornness and stupidity. I lost a few to death, including the best friend I ever had. That was Milton. I've kept a few friends who've lasted, no matter what.

"Y'all want to make the rub?" Rick asked.

"The rub." The girls were giggling all over the kitchen. "Let's make the rub."

Nobody knew where anything was. Rick wandered from shelf to drawer, searching for dish towels, whisks, and spices. Lowry found measuring spoons. The girls meted out cumin, paprika, kosher salt, black pepper, brown sugar. The rub was a potent reddish powder, mixed in a little yellow bowl by Lowry's coquina-like hand.

"Shakespeare's Rub," I said, thinking, *To sleep—perchance to dream: ay, there's the rub!* Mary Katherine lifted her eyebrows at me.

Rick put the game hens to thaw in a sink of water. When he moved away, pink crab legs jutted from the second sink.

"What?" Mary Katherine exclaimed. She pulled a crab leg from the water and tapped it across the table.

"Crabs in Montana?" God knows what he paid for them.

"I found a good recipe for crab cakes," Rick apologized again. I'd never made crab cakes and I guessed Rick hadn't either.

"I dreamed about them last night," he said.

"I swear, I have never dreamed about crab cakes," I said.

We were in much deeper than I'd thought, which Rick was realizing because he said right then, "The hens need a stuffing too."

I didn't want to hear anything else. When I'd seen the crab I'd known for sure there'd be no steep climb to Grizzly Point. The trail up would be scattered with fir cones and crossed by the routes of bears. The view from the top would be smoky with mist, like a magic trick, all the way to the Canadian Rockies.

Mary Katherine tapped over with the crab leg and play-bit me.

"Ouch! Quit being crabby!" I said. She rung her bells. This was the funniest day in the world to the children. Their dad—man of public meetings and fax machines, of manuscripts and manila envelopes—was wearing an apron. He had red powder on his chin.

The recipe for the stuffing listed three ingredients: chopped onion, chopped orange, and chopped star anise. That seemed simple enough.

If star anise meant the dried, star-shaped seed head, chopping them would be like hacking kindling. "Do we have three star anise?" I asked. Rick pointed to plant material wrapped in a plastic bag on the counter. I took it out and examined a bulbous stalk with layers, like enlarged celery. I'd never had any dealings with star anise.

"It calls for three?" Rick asked. "I only got one."

"How far back to the grocery store?"

"An hour one way."

"One's enough," I said, rinsing the anise. This looked like a vegetable, not like an herb. Was it fennel? If the flavor we sought was licorice, wouldn't fennel supply it? "I'm surprised you didn't have to go to Seattle to shop," I said, pulling apart the anise layers, a radical fragrance rising up and out. "It says dice. How big do you think a 'dice' is?"

"Smaller than a dice, I'd say," Rick laughed. "Maybe like an aspirin?" Metaphors and similes were one reason I like to be around Rick. He turns out metaphors like other people turn out platitudes. What writer wouldn't want a friend like that?

Rick was cracking crab legs with a pair of pliers, tugging meat into a bowl. He flipped back to the crab cake recipe—the one he dreamed about—and I flipped forth to the stuffing; we used a chili powder bottle to mark the other's place. I chopped the husks of the anise or fennel or whatever it was.

By then we had climbed many mountains together, Rick and I and whoever else wanted to go, him always ahead, I trying to keep up, my heart pounding like a V8 but never pausing, determined not to lag, making the same leaps to the same rocks crossing mountain streams, sometimes using a tree to push off and sometimes my hands on my burning thighs helping my legs ascend rocky trails, zigzagging uncertain scree slopes, heady with elevation.

After a while we would stop. We would reach the top of some mountain, Mt. Henry or Rock Candy, and some view that Rick had in mind to show us, a view of the Selkirks, peak after peak covered with hemlock and fir and larch, mountains as far as a person could see, and I'd stand with my lungs burning, thinking about all that wilderness around us and what was in it, lynx and grizzly and mountain lion and grouse. Even if we hadn't seen any of that ourselves, it was all in there, in that vast botanical sea. Far below a river was raging.

Rick would drift away around some boulders, and I'd see that he was gathering something and it was wood, and soon he'd have a little fire going in some notch out of the wind, nothing big—a few sticks at a time burning—and to say it was comforting to sit on a rock near this little fire way up in the mountains would not do justice to the moment. It was primal. It was elemental. It was electric.

I'd have so much oxygen pumping. It would be hitting everywhere in my body, and awash in all that oxygen I'd feel as alive as I've ever felt. The rocks were more rocklike, the tamaracks more alive, the clouds more psychedelic than any other clouds. That is the truth.

Then Rick would pull a plastic bag from a pocket and it would be a clump of marshmallows, and he would offer them around and already have some little sticks ready. That's the kind of guy he is, or the kind of friend he is. He would hike four miles straight up to a wind-pummeled bald with half a bag of mashed marshmallows in his pocket and then whip them out. He could keep a surprise.

He had a big heart and it was full of marshmallows on sapling sticks. Each of his letters, each conversation, was a gift. The advice he gave me about how to get something published, or how to edit something so it would publish, was a gift. The courage he demonstrated in the hard fights to save more wild places and wild things, a gift. The strength he offered when I thought I'd lose my own fights, gifts. The inspiration I continued to find in his books and essays when I had none. The jubilance that attended the news of success.

Friendships between men and women carry all sorts of freight. Some people say they're impossible. I was a woman and he was a man, but he had his family and I had mine and we left it at that. I'm not saying that I didn't have a crush on him. I did. I'm not saying that our differences didn't surface. I'm sure he wished I could hike farther faster. He probably wished that I liked to hunt. But he never said this to me. I wished he'd go slower. I wished he liked to garden. I never said that.

We sat up in the mountains not talking much but looking around and smiling some, our brains juiced from the climb and thinking about going back down, how we'd get to fly. I'd gaze across to another peak in the Selkirks, and I'd feel so tiny in all that wildness, in all that rock, and the fire ever smaller. But powerful too.

The velvety marshmallows would come from the smoke burnt and bubbling, their skins caramelized like crème brûlée, sagging with melted sugar.

A friendship is a lot like a fire. It can be hard to get started. After that it needs tending. In my lifetime, I've had lots of friends, probably hundreds, but only a few truly great ones. A person can go around starting minuscule blazes, racing back and forth trying to feed them all; or a person can keep a few atomic bonfires burning high and bright.

The harder I worked at being a good writer, the fewer friendships I could maintain. Writing took a lot of time, and it took a lot of me; and when it was done there wasn't much left of either to share.

We carry so many expectations of friendship. We've heard James Taylor sing "You've got a friend," and we want ours to come running. We want them to work miracles. We want them not to disappoint.

Let's face it. The roadless areas of the Yaak Valley of Montana, wildest place left, were not protected. Something in that disappointment colored my friendship with Rick—if only I worked harder, rose earlier, wrote more letters. I'm not saying that he had these expectations of me. I had them of myself.

I released the idea of seeing the lodgepole pine and old-growth larch on the way to Fish Lakes. When I finished the stuffing, the hens weren't thawed. I sneaked a look at Rick's crab cake recipe, and it was actually four recipes in one. First he had to make crumbs, and for this he needed ten slices of white bread.

"White bread?" We were whole-grain kind of people.

Rick laughed loud and sudden. "The store had two brands, one called Hillbilly. I liked the name but I didn't buy it."

"Why not?" I asked.

"It wasn't white enough," he said. What he bought was Wonder Bread.

Grind the bread, the book said.

That stumped us. Grind the bread. Mary Katherine laughed until she rolled on the floor. She was an incredible laugher, an infectious laugher, better than a flock of flickers. We found a food processor in the pantry; it seemed the sort of tool chefs might use to grind bread, but none of us knew how to operate it. Rick checked the blender.

"It has a button marked 'Grind,'" he said.

"Seems like a sign," I said.

Now the kitchen was a dam-burst of smells, bergamot and anise and onions and something fishy, the yeast of Wonder Bread ground to nibs. Mary Katherine wandered off to a book. Lowry was dressing a doll. I was chopping more onion.

"These hens will make everybody sad," I said.

"Are you crying?" Rick asked.

It could have been the onion or it could have been something else. I was just thinking how much I admired this human, this friend—he was giving his life away to a place and to people he loved, mostly to wildness. Day after day he woke up and tried to make the world wilder and richer and more fun. He never stopped, like wind wearing away rock, getting down to something indispensable. I didn't simply admire him—I loved him—wanted a blazing friendship with him—and I was thinking how tenuous and fragile life is. How nothing is ever secure. "I'd sure hate to lose you as a friend," I said. What was I trying to say? Was I trying to lessen the miles between Georgia and Montana? Was I trying to make permanent something that cannot be fixed?

Rick didn't reply.

But I had come. I'd taken time away from my work and so had he. Like a life, friendship can be defined by length of time or by depth of experience. So I soldiered on. "I'd like to know that, no matter what, for a lifetime we'll be friends," I said. "I'd like to know our friendship will outlast success and failure. Even if it means we don't visit often or talk much."

"That seems a given," he said. "Can you beat these eggs?"

Friendship isn't a given. It's a snail shell, a cloud formation, a zephyr. It's a bird hatching, it's a hummingbird at a columbine. It's a lost amethyst.

We needed a mayonnaise for the crab cakes. The mayonnaise started with eggs and required steak sauce and Worcestershire and

Dijon mustard and other things. While all that was spinning in the blender I dribbled in five tablespoons of olive oil. Mayonnaise flew everywhere.

Mary Katherine rushed back.

"What is that?" she asked.

"Mayo."

"It looks nothing like mayonnaise."

We had been in the kitchen for hours when Rick disappeared outside to build a real fire under a grill where he would lay the Cornish hens. He was gone a long time. When the girls and I went out to find him, he was grilling elk tenderloin.

During Rick's life in the Yaak, wildest of American places, last vestige of frontier living, he had fed his family the pure meat of elk that browsed wild blueberries and of deer that ate lilies and violets and whose meat carried the coinage of flowers. Late last October, Rick had told me, he was in a roadless wilderness when an elk came to him. He was five days hunting it, two days packing it out, seven days aging the quarters in the cool of the garage (frost, each morning), then three days butchering. "But eighteen month's worth of meat," he said, "from those seventeen days." Talk about slow food.

On the same fire he grilled mushrooms. "The morels came from the twin-humped mountains there," he said, pointing. "Every night last summer we watched the fires up there. The mountain was lit up like a jack-o-lantern."

Finally, after all the rigorous tabulation, the direction-following, the mixing, the culinary dalliance, after all the exotic tastes and smells, finally we have arrived at the heart of the menu that Rick has planned, or dreamed, which is the food that came from this valley, that he has gathered. This is what he had in mind all along, another, more direct way of filling us with the place he loved, of making it a part of us.

We were shooting to eat at six, when Elizabeth and her mom were expected. The fare seemed to lean heavily toward meat, so toward the end Rick steamed a bundle of asparagus. We set the table with fine china, even wine glasses (filled with juice) for the girls, and Mary Katherine used a crayon to print names on place cards. The women arrived, all smiles. We brought the feast to the table—the green salad, the painstaking crab cakes with their homemade mayonnaise, the chile-crusted grilled hens, the woodsy wilderness elk, the mouthwatering morels, the smoking asparagus, the good bread. Water and wine. Nobody quite believed it.

I said friendship isn't a given. But maybe it is. Maybe it's something worlds beyond our control or our understanding. Maybe it always circles us, high in the star-filled sky.

As we dined, not far away loons floated and beavers gnawed. Arrowleaf balsamroot and beargrass brushed against the sides of Mt. Henry. Thrushes filled the windows with music.

In the Elkhorn

IN MEMORY OF MICHAEL WINSOR

Helena National Forest, Montana

Townsend, Montana, itself had shocked me with its remoteness. I had fifteen miles to go before I reached a cabin the US Forest Service had granted me after I applied to be writer-in-residence of the wild. The roads gradually worsened amid a vastness of inhuman country as I crept deeper into the Elkhorn Mountains.

Finally I reached Eagle Guard Station, constructed in 1895 of hewn, chinked logs in the high grasslands of west central Montana, and I parked beside a corral that surrounded it. The cabin had a small porch, two front doors, and pretty, white-frame windows. I had been sent a key to the door, but when I climbed the rock steps, the lock on one door had been cut and the doorknob on the other broken. A footprint of a large boot was limned in the dust.

The dim cabin was so richly brown it brought to mind an old hollow tree. A woodstove stood in one corner, a handmade table in another, bunk beds and a bench in a second room. Lanterns that the Forest Service promised were gone, along with the propane stove, pots, pans, and shovel. Buckets for hauling water were gone.

I retrieved a cowboyesque revolver out of my truck. My father taught me to shoot when I was a child, and although I hadn't shot in years, keeping a gun could bring gravitas and bravado to a situation.

So I kept one. Now light rain was falling. I balanced a lumber scrap on a post and popped at it with the .22. On the fourth bullet the scrap fell.

I found two six-inch nails and nailed the bunkroom door shut. The front door bolted from the inside.

Then I kindled a fire in the stove. At 6,800 feet, the weather was cold for June, down in the 40s. I needed water, so I found an empty gallon-sized coffee can and headed down-canyon, through an idyllic meadow of lush fescue, tufted hairgrass, long-plumed avens, and cranesbill. Eagle Creek was two feet wide. Mountain bluebells hung over the mineral water, and above it, quaking aspens shook and shimmied. If cougar were anywhere, they were here. I filled the coffee can.

The rain stopped while I ate supper, and a setting sun poured brilliant fire across the gilded mountains. A rainbow commenced in a far canyon, growing more luminous as it arched and touched down so close, just up the creek. In minutes a second rainbow materialized from out of a cloudbank and bent over the first. I sat electrified as dark descended, watching arches of spectral light and the last rays of sunset stroking a leaf-line of peaks.

I was there, way out there, alone. I was somewhere in history, alone.

I had invited my friend Michael to come. I collect interesting, complicated friends, and he was at the top of the list. Thin and wiry, his eyes were warm but always held something back. He was a painter in the art department, exactly my age, also a graduate student. He mostly painted seeds. Sometimes I went and sat on a ratty sofa in the room where he and other art students painted. His studio was in a back corner. He stuck Van Morrison on a tape recorder and stood at his easel, drawing huge black scribbles of seeds amid a flurry of red sky. But he wouldn't join me in the wilderness. I knew the idea had scared him.

Next morning I descended again to the creek, noting a zone of sagebrush, where sage sparrows and sage grouse frequented, into the juniper and Douglas fir of the canyon. I came upon a mule deer kill. Hair and hide were strewn about, bones stripped of meat but not yet porous and bleached. I poked through the putrid carcass and uncovered the skull, intact, which I toted back to the cabin, feeling nauseous.

Inside the corral, death-camas bloomed, and mountain iris with its poison seeds and roots.

Tree swallows were nesting in a crude bluebird box. I read in the cabin journal that a father and daughter, previous guests, had built it the previous winter. They asked in the journal if someone in the fall would please clean out the nest for the next spring's birds. That would not be me, because this was summer and the box was in use. I decided to leave them a note in case they returned. "I wanted to let you know that your birdhouse is being used by swallows. The bluebirds you meant to shelter are actually nesting in the swallow's mud cup under the eaves!" The female bluebird would fly toward the cabin window and drop again and again before her reflection, catching herself inches off the dew-tipped grasses.

I rambled out into the mountains, watching for sign of mountain lion, and when I reached larger Eureka Creek, I walked along it scrutinizing animal tracks. I recited poetry to myself, including H.D.:

> O be swift.
> We have always known you wanted us.

In the afternoon I sat in the sun with words and ideas, which are lifelines. I drank chicory tea.

Everything was existential.

When I broke from writing I chopped wood—the axe had not been stolen—and fetched water. I washed dishes. I rinsed out a shirt and pair of socks and hung them on a rope to dry. In the evening I walked again for an hour or two, following scant trails, and sometimes I entered the canyons. Late candles guttered over the pages of books.

The hours like crippled dogs dragged themselves along.

Months later, after the tragedy, I retraced my steps through the Elkhorn, when I studied life and death.

Earlier in the summer, a strange thing had happened. I had been on the Missouri River, canoeing for five days with another friend. Mick was a tall, strong Englishman who'd been a fell-runner in Scotland. He was immensely capable, and we'd taken a number of backcountry trips together. In fact, in my front yard on Chestnut Street in Missoula he had built the cedar-strip canoe we were in. The river was swollen, two feet above flood, rising into beaver lodges and gnawing at cutbanks as it swept us along. We kept midstream to avoid crashing wedges of earth that would sink a boat with no place to climb out. The river was a muddy brown, to the point that any life contained in it was invisible.

On the fourth afternoon we had stopped on Holmes Council Island. The plains sun was gentle and there was a wind. I was lying against a fallen cottonwood looking up at thousands of cottonwood hearts throbbing in wind, hanging by their heart-strings. They were like hummingbird wings fanning tiny flames. I was warm and safe against the belly of earth. But something in the river's motion, thirty-six miles of it so far, had entered my body, and I found myself rocking, tipping, and correcting. This had a startling effect. Even on dry ground I was adrift.

Suddenly I felt as if my soul were leaving my body, rising above it into the sky. That's how it felt, as if I were outside myself. The whole of the world—the breaks, the wide, muddy river with its animal-like lapping, the incessant wind, the sun in the dry blue sky—began to warp and slide away. Or maybe I was sliding backward, the land-scape melding to a green and blue distortion. I was neither in my body nor of the earth, but disembodied and helpless, as if somehow I had slipped outside myself while watching the cottonwood leaves, creeping out with steady exhalations, but had not gone far, not out of camp. My soul was hanging close to my body. But not inside it. Definitely not inside it.

My entire body was like a slept-upon arm that has become numb. With an arm, you wonder if it will ever work again, or if it will forever be as worthless as a wooden spoon dangling from your shoulder. Except now my whole body was numb. I could just as well be dead. Maybe I was dead.

I sat up.

So I could move. If I could function, spirit had not completely abandoned my body, but somehow pulled it along, a sled hitched to a line of running huskies.

My friend hunched in the grass some distance away, reading.

"Mick," I said. I could still talk! He looked up and put his book on the grass.

"I feel odd," I managed to say. Tears flew out of my eyes. "My body is leaving the earth." I tried to focus on the worn rock of the cliff, to see it apart, in its place, not part of me or me part of it. Then my friend was on his long legs, coming toward me, bending, clamping his arms around my shoulders, holding me down.

I bring that story up to say how untethered I sometimes feel.

I have a memory of walking with Michael in the woods above Mt. Sentinel in Missoula, not far from campus. Birds fluttered in the green treetops. There, not far above our heads, was a pair of the most beautiful lovebirds I've ever seen. The Western tanagers seemed unreal, birds painted on canvas. They seemed like messengers.

As the days passed at Elk Guard Station I didn't touch another human, didn't speak to anyone. There was no one to see. In the mornings I hiked out through grassy knolls, into the mystic. The trails weren't to be trusted. Through the forests they were blazed, and traceable, but in the wide meadows, paths disappeared into lupine, arrowleaf balsamroot, and gromwell. Keeping to them took concentration.

Midweek I began to fast, ascetically, stripping more of the world away to get a good look at what was left behind. One morning I trekked toward Peck Mine, six miles away along a rutted Forest Service road. I was close to "zero at the bone." My dreams had been ragged, Michael with his back to me. I pulled my body along, unable to walk far without resting, and my heart beat hard in my shrinking chest. I felt it pumping. Even the air was lean.

Two yearling mule deer stood alert on a near slope. They seemed torn. One watched me while the other turned away, distracted by something else. Just then a coyote popped up on a nearby hill, at first trotting, then galloping toward an outcrop of gray rock where a few Ponderosa pines grew. A third mule deer was chasing it. The coyote circled out of sight, then back, trotting beneath low-hanging pine branches. I assumed that under the pine the mule deer could close in but not thrash him. From under the boughs the coyote watched me, watched the deer, and finally, uncomfortably, jogged into the open. The mule deer was again in pursuit. They soon disappeared.

One afternoon I lay outside beneath a strict sun, reading. A golden eagle circled above me, lowering until it was less than a hundred feet overhead. As it lowered it grew, until it was immense, a mythic vulture, a winged reaper, capable of carrying me away. I leaped to my feet.

In the Elkhorn, every direction of the compass pointed to death. Bones littered the yellow-green hills, jawbones of mule deer, elk femurs, deer pelvises with their gaping holes. When I picked up a two-foot length of backbone, its vertebrae fell apart in my hands. Life is fragile, breath and blood and white bone. Life is a cairn in a high wind. Life is surviving a chase. Life is war, ancient and long.

I lived in an apartment across the hall from Michael's. Separated by a thin partition, I could hear when he came home, when he pulled his Murphy bed from the wall. Toward the end of summer, Michael took a long trip to the West Coast. He left in a hurry one morning before day. I heard him go, taping a note to my door, which asked me to water his plants. When he came back, he brought a bowl of fat Washington blackberries, and I came over to his apartment and we ate them, sitting in straight-backed chairs.

That apartment is where Michael hanged himself. But this was later. A few months had passed. I had graduated and left Missoula by then. Friends called to tell me.

A mockingbird began to come and scratch at the windows of my house.

I knew it was not Michael. Not really.

On the last full day in the Elkhorns, following a topo map, I bush-whacked to the confluence of Eureka and Crow Creeks. Trailblaz-ing was hard, through willow and alder, but I was eager to see a place I'd read about, Crow Creek Falls. I collected moose pellets, thinking about Chuck Jonkel, a biologist who showed me they burn like incense. They're dried herbs and bark, after all.

I forded Crow Creek, hiking boots around my neck, to gain a foot-trail I could see on my map. I found it, a path well-worn and unspeakably lovely. It followed the creek past talus slopes and cedar trees, and I didn't put my boots on right away.

Two hours later I reached the falls, startled to find old mining equipment that had been heli-lowered into the crevasse and never removed. The creek banks were littered with rusted cable and plastic bottles emptied of petroleum.

But the waterfall was spectacular. Its water did not fall but hit the tripwire of narrow gorge and exploded into a swirling pool fifteen feet below, releasing the odor of ozone and chicory. A few tree trunks were caught in the pool, where they spun like bottles. The scene made me feel flimsy.

After a while I headed back, returning to the cabin by a different trail that promised the guard station in four miles. It climbed a canyon where Tom Brown Jr. could write a thriller from the tracks: elk, coyote, mule deer. I could easily have been the first human on this trail in a year, two years. It was faint, and I didn't like feeling so lonesome and rickety on these whimpers of trails.

After a mile I lost the path and instead followed my compass east. I left a canyon and traversed a meadow, then linked onto a scrap of trail. I stuck to meadows, trailing east, and bisected canyons when I had to. The trail was gone. There were no signs, no blazes. Overhead a thunderstorm brewed, clouds racing like fast cars above me, until a guild of thunderheads eclipsed the sun. I had begun walking faster and faster, until now I was running. I crossed one meadow, wove through a patch of pine and Doug fir, and entered another meadow. Sometimes I would get to a precipice and have to scout and retrace, then take another route. At that point only God knew where I was.

A storm—a bad one—was imminent. In times of panic I say, *To hell with positive thinking.* I prepare my mind for the worst. As I ran, gripping my pack to keep the pistol from thudding against my back, I considered what I would do if my running attracted a mountain lion or if I was forced to spend the night in a cold mountain meadow.

On the next high knoll I paused, panting. There I recognized the outline of a certain peak visible from the cabin, then the familiar silhouette of a ridge I'd climbed a couple days earlier. The ridge was still some distance away. The storm now blotted out most available light and I could barely read the topo map, so I stopped long enough to hold the compass needle still and memorize the topography before me. I had a meadow to cross, then Eureka Creek canyon, then open prairie again, maybe a mile to go. Also, if I could find Eureka, I could trace it home, even in the dark. And if I couldn't, well, I had a rainsuit and a gun, a knife and a compass. I had my head about me. I had my soul.

The clouds hung low and livid, the sky a blue and darkening bruise. I angled downhill, tripped on a stone, and fell. My right cheek throbbed, and when I touched it, my hand came away bloody.

In the middle of the high, wide meadow the lightning started, slicing down with electric knives, railing against the scabbed earth. Lightning ripped the clouds open, loosening rain and mixing it with pebbles of hail, which hurled coldly against the ground.

Hastening to the meadow's edge, I took refuge near a grove of pines. I needed to keep going. I shifted direction, skirting the moorland, headed obliquely where I thought I needed to go. The rain was hard and lightning crashed at the same time thunder struck. But the compass's red arrow in my memory was guide enough.

I jogged on through the dusk, soaked now, passing out of the storm and setting on a trail at Eureka, looping along a logging road I knew, rising into the clearing of the cabin as the day disappeared. Like so many travelers, I'd come uncountable hard miles and had also seen unaccountable beauty.

When the week was done, I drove straight to Michael's house. In the afternoon quiet of the Sunday town, he was napping.

"You should have come," I said.

"I couldn't," he said.

"I wish you could have."

The last morning, I told him, there were crossbills in the creek. I'd hidden behind a gooseberry and watched them bathe where the creek ran shallow.

This was the first time I'd seen crossbills, the charcoal X's of their beaks. The sun was young and bright. Pine siskins came to join the crossbills. The birds, red males and yellowish females, slung water and chirped, dipping and flinging. They exhibited little fear, so great was their joy to be wading in the half-inch of clear water in a sand bed of a summer creek. The siskins twittered among cress.

I was less than six feet away, I told Michael, part of the world and glad.

PART II **MIGRATION**

The Duende of Cabo Blanco

A LOVE STORY

Cabo Blanco, Costa Rica

A thin man in boots was bending over a sawhorse when I stepped into a compound at the southern tip of the Nicoya Peninsula, western Costa Rica. I had come from the road through a footpath about the width of a machete, arriving at Cabo Blanco, where I would be spending a month. A simple wood-and-screen building, like a feeble citadel amid the green and strangling verdure, perched atop a cliff that overlooked the Pacific. I could hear the ocean murmuring.

The thin man at the sawhorse sensed my presence and looked up, and so I waved and headed toward him, preparing to launch my practiced speech. It was then that I noticed some class of small and menacing dinosaur creeping up behind the man. The man had not seen it. My mouth fell open and a word tried to fall out, except I had no word for this thing.

The thin man jerked around. "Oh-ho-ho," he laughed, half to himself. "Es un gobarto."

So this was an iguana. For a lizard it was immense, four feet long. It crept along, its skin green as oxidized copper, color of weathered Spanish helmets, and it lifted its intimidating head a foot into the air at me. It had beady eyes with a narrow look in them. It had a triangle for a mouth, jaws no doubt lined with razors.

"It won't hurt you," the man said in Spanish. "Are you a hiker?"

Now the iguana was creeping toward the forest, which pressed in from all sides. "I'm a volunteer," I said.

"This way."

The door to the veranda was screen, the floor made of concrete painted barn red. The man indicated that I should set down my backpack.

"María?" he called. Inside, the house floor was the same red, the walls white and papered with posters. I could see a living area, bookshelves, and a television.

A young woman answered cheerily from the back. She was petite and pretty but not in a soap opera way, with wavy black hair and a very happy countenance. "A volunteer," the man said to her and went out again.

"I was making tortillas," María said. "Just a minute." She washed her hands and told me to follow her to another doorway. "Women sleep in this room," she said. "Pick any bed. Unpack. Lunch will be ready soon."

I unpacked and walked around reading the posters, in Spanish— save the owls, don't eat turtle eggs, Chief Seattle's brotherhood speech, acupressure points. They were reassuring. At lunch, three of us—María and I and the thin man, named Alphonso—sat down to a meal of beans, rice, and salad. They asked me questions that I tried to understand and answer. How long would I be staying? A month. Where was I from? Florida (at the time). What was it like living in Florida? Nice. Was I a scientist? No. Did I like nature? Very much. Was I married? No, I wasn't married.

I asked them questions. Nobody said a word about *duende*. Or *jefe*.

In the afternoon María gave me a tour of the compound, which consisted of picnic tables, a bench for sitting and looking out over the ocean, a tool shed, and a three-sided hut with a thatched roof

that she called the museum. My first job, María said, was to clean
the museum.

Cabo Blanco Reserva Natural Absoluta is Costa Rica's oldest national park, established in 1963. About 3,000 acres, it's shaped like
a triangle and bounded on two sides by the Pacific. A fifty-shades-
of-green forest meets the smoky, restless ocean along a rocky,
driftwood-strewn beach. The park is named for a tiny island, Cabo
Blanco, rising off the point, because during the dry season its rocks
are stained with seabird guano, thus "white end."

I'd been trying to understand the word duende. *Literally translated,
it means "boss,"* jefe, *but the word also means "spirit" or "soul," as well
as an elf found in Spanish mythology. It means "elusive aesthetic" or
"transformative power" or "magneto." Poetry could have duende, as
could flamenco music or any art. But not all art had it. Duende was
what was real in art. It was mystery, struggle, darkness.*

The rough shelves of the museum, open on one side to sun, wind,
and rain, were arranged with whale bones, shark ribs, and all manner of natural ephemera, dusty and cobwebby but obviously collected with great care, even love. I picked up the jawbone of a dol-
phin, brushed it, and replaced it beside a curling label that offered
facts about dolphins in multiple languages. I held ribs of sharks,
carapaces of land turtles, skulls of sea turtles, vertebrae of iguana,
teeth of monkeys. I picked up shells in the forms of spirals, scoops,
scalpels, and dots, as well as intricate corals and sponges. I blew
dust from the outspread wings of blue morphos and other tropical
butterflies pinned to a shank of board. Amazing things bloomed
inside jars of formaldehyde, oddities like jellyfish, eels, marlin eyeballs, and baby squid. I dusted the tender nests of birds, wasps,
mice. I wiped laminated sheets of flora, back and front.

In the evening I went to sit with María when she spread an oilcloth on a picnic table and sorted dry red beans, tossing the wrinkled, brown, or split ones. From the lawn, the ocean seemed endless,
water layered gray to green to blue across the patina of the world.
Far far away, a strip of the faintest violet gauzed the horizon. As we
sat the sound of buzzing insects bloomed from the trees. "It's ten
minutes to six," María said. "You can set your watch by the locusts."

We humans live within rings of sound. At the hub are the click-
ing of bones and sinew inside our own bodies and the sibilance of
our breathing. Then come seams of conversation, María speaking
slowly to accommodate my kindergarten Spanish, followed by my
tentative responses. The lines of conversation sew together the long
fabrics of quieter sound. Beyond our dialogue the beans scuffed
along the oilcloth; the locusts wove a diaphanous tapestry around
us, their buzzes rising and falling, accompanied by the gongs and
squawks of birds. Farther out, the ocean moved in flux and reflux,
roaring, grinding against its stones. Beyond it all, there was the
eerie, whale-like undertone of the universe singing to itself.

María finished sorting, folded up her oilcloth, and went inside.
I heard a snap overhead. A monkey crouched on a branch, watching me as it chewed a leafy twig. I saw another, and another. One
carried an infant, clinging to her back, and another young monkey
shadowed its mother. A troop of howlers (*congos* in Spanish) were
feeding, a ladder's height above my head. I could easily have climbed
to them. Marie told me later that the monkeys sometimes piss on
sightseers' heads.

*If duende is what is real, if it is power and magneto, if it is
mystery, then the monkeys had duende, the museum had duende,
the seething ocean had duende. The beans on María's oilcloth had
duende.*

That first week I saw little beyond the compound. I got to know the park guards who lived in shifts away from their families in Cabuya, a village a few kilometers away of about 300 people, mostly subsistence farmers and fishers. Whoever was on duty would eat lunch with us. The boss was away, María told me.

"There is a boss?"

"Yes. Joaquín."

"Where is he?"

He was at San Miguel entertaining a contingent of Danish supporters. San Miguel was a scientific station on the other side of the reserve. It had a nice beach.

"When is he coming back?"

María shrugged. "Soon," she said.

I learned to register hikers as they arrived. One of my jobs was to search their backpacks and take away their knives, so they wouldn't be tempted to carve flamboyant advertisements of passion into trees. At the end of the day I'd return their knives. The rest of the time I helped Alphonso build a storeroom. Day after day I worked painting it. When that job was finished I cleaned, organized, raked, collected trash, cleared brush, repaired nearby trails, painted trail signs, hauled supplies, and helped María wash dishes.

After the first week I began to walk to the village almost every evening. One store operated out of the front room of the house of a crinkle-eyed woman named Rosa Aura. She sold white rice, red beans, cornmeal, beer, and small packets of cookies in her *pulpería*. The only public telephone in Cabuya hung in Rosa Aura's living room, sixty *colons* per call, less than a dollar.

I went to the village to call my son. He was only two, home in Florida with his father, from whom I was divorced. This was the secret I bore, that I was split in two. My son could barely talk. His father would hand him the phone, and he would make noises and some would sound like words—he knew my voice, of course, and knew I was somewhere alive in the world, even if I was nowhere to be found—but he couldn't stay interested in a phone except as the object that it is.

Most mothers of a two-year-old would not leave a child—a beloved child—for a month. I want to explain myself. I had unexpectedly become pregnant while dating an older man before I realized how mismatched we were. In our idealism and our desire to become a family, not yet knowing that would be emphatically impossible, we had decided to marry. I told myself that I could pursue my dreams although I was a mother. I could travel. We'd figure out a way to do it. My son has become a beautiful human being—he has a kind and understanding nature and makes decisions based on his heart, not his head. I have never seen him hurt another creature, ever. That is now.

Then, he was two and I was determined not to let motherhood get in my way. So I was volunteering at a national park in Costa Rica. Besides a train trip to Canada, it was my first trip outside the country. I missed Silas terribly because a month is a long time, and sometimes I cried. That too was mystery, struggle, dark sounds.

The first time this happened, I had hung up the phone after a completely unsatisfying call. Was he being cared for with the attention and love I would give him? Was he wracked with longing for me? Was he forgetting me?

Rosa Aura wanted to know what was wrong. After that there was not a soul in Cabuya who wasn't friendly. An evening crowd would gather in the store, making jokes and laughing. They called me *Mamita*.

I would walk back to the bunkhouse in the half-dark in a somber mood, thinking about my life, how I was going to navigate it now that I was a mom, with dreams and responsibilities in equal measure, and thinking about Cabuyan lives, how in ways I wanted to live as they lived and how in ways I did not.

Thus I fell into a pattern of days. When I had time to help María in the kitchen, she taught me to make tortillas, patting out balls of corn dough and flipping them on a hot griddle, pressing them with an oiled napkin on the third turn so they began to fill with air and rise.

"What do you write in your book?" she asked me.

"I try to write poetry," I said.

"El Indio likes poetry," she laughed.

"El Indio?"

"That's what people call Joaquín."

"To his face?"

"Behind his back."

The Indian. "Why?"

"He looks like one. You'll see."

Toward the end of the week María asked if I'd help her paint the kitchen.

"Of course."

"We have to hurry," she said. "Joaquín comes day after tomorrow."

"Why does that matter?"

"I have not asked permission to paint," she said.

"He'll be happy you painted," I said.

"You don't know Joaquín," she said.

All day we whitewashed walls, using the last gallon of latex, and at dark I fell into my bunk aching. María had remained nervous, because women were lined up to take her job, I figured. Her nervousness was contagious. The following morning we were up early again, scurrying to replace pots and pans, dishes and food.

"Is he that bad?"

"Yes." María told me stories of Joaquín's acrid tongue, the laws he laid down to worker and tourist, the guests he evicted. "Watch out for him," she warned. "He likes women."

"He's not married?"

"No."

"How old is he?"

"Thirty-six." She glanced conspiratorially at me and repeated herself. "He likes women volunteers."

"What do you mean?"

"He sleeps with all of them."

"All of them?"

"Almost all."

I remembered then a tall, mustached director in San José, when deciding where to send me, had said "Cabo Blanco." Another heavy-set administrator nodded. "Cabo Blanco." A funny look had passed between them. The next moment I stood in front of a map, memorizing key words. Take the bus to Puntarenas, spend the night. Catch the 6 a.m. ferry across Nicoya Bay, a taxi to Montezuma. From Montezuma get to Cabo Blanco any way you can—perhaps a fisherman will carry you by boat. A jeep goes once a day. Find Joaquín.

They exchanged another look. "Have a great time."

The Indian arrived by land minutes before noon. María was preparing lunch in the sparkling kitchen. She was particular about what she cooked for Joaquín. He was picky, plus a diabetic. That day she had steamed green beans and was wrapping them in a tiny pool of whipped egg, small omelets. She heard a noise and checked the window. Her spatula hit the pan with a clang.

"He's here," she said.

A man stepped in the back door. He was a dark-skinned, barrel-chested man dressed in shorts who reminded me instantly of a mountain goat, maybe because of his thin ankles. He wore calf-high leather boots tightly laced. A yellow band held back long, raven-colored hair, and a clutch of talismans hung from his neck, cords tangling on his chest.

María concentrated on the stove. "Hola," she said. Joaquín looked around.

"Qué pasó?"

"We painted the kitchen," María said, glancing my way. I was tight beside her.

I smiled.

Joaquín grinned and María fell against me, laughing.

Joaquín looked around again and the smile left his face. "Where are my posters?" he said. We'd left a dozen faded and dogeared posters off the kitchen's one blank wall.

"They were old and ugly," I began.

"Did you throw them away?" That was the only thing he'd said to me.

"No," María said.

"Put them back up," Joaquín barked. "All of them." He stalked out.

After lunch, once I'd taped up the ratty posters, I went back to painting the baseboards in the main room a chocolate brown, and the next morning I began painting the porch walls, thinning the color with kerosene.

Armando, a paid guide who came almost daily with guests, arrived. "You must like to paint," he said. Joaquín was watching me with the guests.

"Not really," I said to Armando.

At ten Joaquín abruptly asked me to stop. "Vamos al bosque," he said. Let's go to the forest.

I stepped through a doorway and looked at María with an eyebrow lifted. She had heard. She rolled her eyes.

Spanish poet Federico García Lorca wrote, "The duende, then, is a power, not a work. It is a struggle, not a thought. I have heard an old maestro of the guitar say, 'The duende is not in the throat; the duende

*climbs up inside you, from the soles of the feet.' Meaning this: it is
not a question of ability, but of true, living style, of blood, of the most
ancient culture, of spontaneous creation." Lorca wrote that Goethe
had inadvertently defined duende when he wrote about "a mysterious
power that everyone feels but that no philosopher has explained."*

Once we entered the forest Joaquín turned and laid a finger against
his lips. After some meters he stopped and made a pawing motion
with one high-booted foot. I raised my eyebrows and lifted a hand,
palm up. Joaquín put two long fingers to the crest of his head. Deer.
Farther down the trail he picked a palm date, turned sideways, pan-
tomimed how to eat it, then handed it to me.

Silence is the resting place of a forest. But a forest is never silent.
Every move makes a sound, and something always moves. The wind
passes underneath a single leaf, gently lifts it, then eases it down.
Earthworms slide through loam, causing grains of dirt to make the
tiniest of clinks as they knock against each other. Sap rises and falls
through xylem. Trees whisper as they grow, pushing upward and
outward, stretching toward the sky. Leaf-cutter ants hurry along
with their ripped green sails whistling.

Perhaps Joaquín wanted to hear that.

Or perhaps he wanted silence because words cause confusion.
Words leave multiple and often incomprehensible meanings dangling all over the vegetation. Or maybe Joaquín wanted silence in
order to lure wildlife, so he could amaze me. That would be seductive.

I walked silently through the jungle along Sendero Central on
the Nicoya Peninsula, in a reserve that Costa Rica had the foresight
to save, following a man who reminded me of a goat. Suddenly
Joaquín spun around, frowning, and put a finger against his brown
lips as if he had read my thoughts, as if my mind was a chatter-
box. He stepped off the path. The many totems hanging from his

dark, corded neck rasped and rattled. In a few meters we reached the largest tree I'd ever seen, so tall its leaves were a green cumulus cloud far overhead, their shapes indistinguishable without binoculars. Its trunk was like a wall.

Then I heard voices. Had Joaquín heard them? Is that why he frowned?

"Shhhhh," Joaquín said aloud, drawing out of sight of the trail, behind the bole. "If people find this tree, that will be the end of it," he whispered.

"How will hikers destroy a tree?" I asked.

"Shhhh."

Three trekkers, a man and two women speaking German, passed. When the forest was quiet again, Joaquín whispered fiercely. "They will tell their friends. More people will come, and more. They will leave their trash. They will carve their names in the bark. They will pack down the earth around the tree until it finally falls over."

"No, they won't," I said brusquely. "They aren't allowed to carry knives."

"They sneak in knives," he said.

"You hate people, don't you?" I said.

Joaquín looked hard at me with his black pebbles of eyes. "Yes. I do."

"Well, I don't," I said. But he had turned abruptly and started off.

Soon he stopped at a vine with holes in its large leaves. "Cobija del pobre," he murmured: blanket of the poor. He stopped again at a vine he called *escalera de mono*, monkey's ladder. He showed me *naked Indian* tree, also nicknamed *gringo's nose* because of its peeling bark. "It's used to treat diabetes," he said.

"Do you use it?"

"How do you know?"

"María told me."

"Yes, I use it," he said. He moved to another tree and handed me a divot of bark, indicating I should pop it into my mouth. "Alligator tree, an anesthetic," he said. He pointed out *caña agria* in bloom, massive pink flowers like cotton candy emerging from a cone-shaped calyx, a diuretic. Ferns enormous as trees towered above.

For a while we tagged a band of white-faced capuchins, *cara blanca*, that swirled through the canopy foraging tender whorls of *pachote* (kapok) leaves. "Taste these," Joaquín said. "You'll see why the monkeys love them." When we came upon a lemon tree, Joaquín pocketed a handful of fragrant fruits. Then, with a turbulent look on his face, he strode on, zesty and paleolithic. Even after we turned homeward Joaquín continued to whisper aloud the names of things, a tropical nomenclature of biodiversity, a luminous glossary, a language poem.

If spirit is what the eye cannot see, then silence is what the ear cannot hear. If noise is superficial, then silence is profundity. Silence is a deeper listening. Silence is shamanic. If some music has duende, so does some silence. There is no primeval silence. All of mystery has a sound. Spirit has sound.

Animosity developed between Joaquín and me. He was imperious and chauvinistic; I was insubordinate and haughty.

One morning I confiscated a jackknife from two tourists. They needed something to spread peanut butter for sandwiches, so I loaned them a butter knife from the kitchen. Joaquín saw it when they returned.

"No knives," he said irritably.

"It's a butter knife," I said. "How were they to make lunch?"

"With a stick," he said.

"You're saying a tree matters more than a person," I said.

"Yes."

"Well, I think you should strike a balance."

"I'm the director of this park," he said harshly. "And I said no knives."

"You're a crazy man."

Still, Joaquín asked me to translate a study regarding carrying capacity for Cabo Blanco, recently completed by two graduate students from the United States. It was in English.

Two volunteers arrived from California. They were surfers, tolerating the work in order to access the ocean. Evenings after I'd walked to the village to place my call and attempt to reach my little boy, Joaquín and I talked about what the study said while María and the Californians watched television. The essential question seemed to be: The park needed the revenue generated by tourists, but did it need the tourists? Trails became bogs and mudslides. Tourists disrupted animal populations. They trampled vegetation. The study suggested the number of tourists in Cabo Blanco be limited since park facilities were not equipped to handle growing numbers.

"More tourists come to Costa Rica every year," Joaquín said one evening. "We have reached the limit this reserve can ecologically handle, and now what do we do? Do we require visitor permits? Do we close the park during the rainy season? Part of the problem is foreign investors." His lip curled at me. "They've bought Montezuma and are coming this way."

Joaquín wanted to double the size of the park. He had gone looking for money in Europe and the United States. He'd made a video, a television special. It was a race against time, his words.

"Why doesn't Costa Rica buy the land?" I asked.

"The park service has no money for land acquisition," he said.

"You want more land and no people."

"People can visit other wild places," he said. "Not Cabo Blanco."

Still, some afternoons Joaquín would say, "Let's go to the forest," and we would stroll in silence through the peaceful and healing jungle. In all those hikes he never propositioned me or touched me in any way.

One day on the rocky shore, hiking back to headquarters, we were caught in a terrific squall, as if God drove a chariot across the skies, ripping open golden clouds, slapping ropes of lighting together to make thunder. Ahead of me, Joaquín raised two mahogany arms to the sky. He turned his face upward. "My god," he yelled into the heavens. "My god." He whirled toward me. "This is my god." Suddenly he went down onto his bare knees in the brown sand, kneeling in frothing surf, long hair clinging to him like bark. "This is my religion," he said.

As Lorca wrote, "The duende...does not appear if it sees no possibility of death, if it does not know that it will haunt death's house, if it is not certain that it can move those branches we all carry, which neither enjoy nor ever will enjoy any solace." In the light of the death of Lorca, murdered during the Fascist takeover of Spain, these words disturb me. I am pretty sure he saw the possibility of his own death. I think he should have taken more care to avoid it.

The afternoon that everything changed, Joaquín invited me swimming. The Pacific Ocean wasn't the limestone sinks and lazy brown rivers of home. The tide was receding, causing a current strong as raw whiskey, one I didn't see or suspect. As we paddled about, whitecaps crashed heavily around us, throwing garlands of surf on our heads.

Suddenly tiring, I tried to stand. I couldn't touch bottom. I began pulling toward shore, which was farther away than I'd realized. The waves wouldn't stop sucking at me. I chose a large black rock and eyed it as I stroked. The rock refused to move. I swam harder and harder, yet the tide pulled me out. Soon I was struggling.

Joaquín appeared beside me, just out of reach.

"Where is bottom?" I gasped.

"Swim," he said, calmly and firmly. "Go. More slowly. Go." It was an order. He meant it. "Go." Joaquín swam nearby, easily. "There is a ledge soon," he said. "Go." He stroked ahead to find it. He was fifteen feet away. I didn't know if I could make it. I was gasping for breath, pulling hard, very tired. "You're close," he said. "Don't give up." I had to live. My life was not done. I had a small child who needed me.

"Here," Joaquín said. "It's here. Keep coming. Don't stop. Swim."

When I was close enough, Joaquín reached out and pulled me to the underwater outcrop. I clawed for the rock with my feet. My limbs were jellied. The ocean released me grudgingly, and I stood weeping in the roller-coaster waves, hugging my chest, shaking, my heart pounding. Joaquín rose like Poseidon on the rock beside me. He put his arms around me.

"Are you okay?" he said.

"I. Almost. Died," I panted. I was quiet for a long time, trying to catch my breath. "Thank you," I finally said.

"De nada," he said quietly.

We still had to reach the beach, and once I was rested we set out, Joaquín paddling an arm's length away. We toweled off silently, dressed silently, climbed the cliff stairs silently. Monkeys chortled in the mangoes while the hostile and streaming Pacific hissed like a vitriolic spirit.

Lorca writes about four elements of duende: irrationality, earthiness, a heightened consciousness of death, and a touch of the diabolical. It was all there at Cabo Blanco, and it was all there in Joaquín, in one package.

San Miguel was marked on the map as an outpost on the western side of Cabo Blanco. Most of the volunteers got to visit it, María told me, but because I'd been the only volunteer for most of my stay, I'd been confined to headquarters. My time at Cabo Blanco passed steadily, as time does.

"I won't be able to see San Miguel, will I?" I asked Joaquín.

"You can see it, if you like."

"I'd like that."

Two days before I was to depart, Joaquín told me to put a pack together. I'd need water, toothbrush, a swimsuit if I wanted, a change of clothes. We set out on foot through the forest. In a matter of hours Joaquín and I walked into a clearing as if it were a moon landing.

San Miguel was like nowhere on earth I've seen. I keep using the word *paradise*, but this actually was pretty close to paradise. A weatherbeaten cabin occupied a black beach, a collection of bones scattered around two fire-rings in its yard. The cabin, crude but homey, with a powdery concrete floor that smelled of prehistory, offered up a Spanish-Danish dictionary, candle stubs stuck in bottles, and altars of shells and driftwood.

Joaquín fired up a generator in a shed for lights. On a propane stove we cooked a soup with meager supplies left by two Australian volunteers. We built a fire in a ring and sat staring into the flames, listening to the Pacific challenging us to be better people, to love one another more, to confer more of ourselves, to not give up.

I don't want to tell the next part, because I don't want María to know that I succumbed like almost all the women volunteers. That night I slept with Joaquín, both of us in a single bed, tongue-tied and alchemical, two humans on a wayward beach that will never be re-created, one to return to a different place and the other to stay in the place he was becoming. That's all I'll say about it. I was a younger woman in a country foreign to me, but he did not take advantage of me. I chose what I chose.

The next morning botanists from the university in San José arrived to gather pachote seeds. Joaquín went off into the forest with them, and I sat by the ocean translating more of the study. I was almost done. *To preserve this delicate balance*, the study read, *we must limit the wearing activity of humans.*

In the afternoon the botanists departed and Joaquín suggested we snorkel. The reefs were alive with brilliant fish: angelfish and butterflyfish, bannerfish and blue tang, peacock grouper and humbug, one humphead parrotfish. A school of powder-blue devil damselfish approached, parted, and surrounded me, curious yet timid. Standing on the dark rocks, one would never know that underwater existed a separate world, illuminated and ethereal.

After dark Joaquín invited me fishing along the rocks. We used a *cuerda*, an atavistic fishing rod that is simply a square of wood with a handhold carved into it. The fishing line wraps around the block. Joaquín showed me how to spin a baited hook across the water, unraveling the line, letting the incoming tides return the bait across the rough rocks.

"You stand here," Joaquín said. "I don't want you to hook me in the dark." He chuckled at his own joke.

I felt a hefty tug. "Fish," I called to Joaquín, who came running down the beach. Just as he reached me there was a snap. Using his flashlight we watched a ghost-white shark cutting away. I stared at the shaft of water visible in the beam of light. Even now below us the darkened waters were filled with jeweled rainbow marvels.

After that I drew in my line and watched Joaquín. He had balanced his light on a rock, and with panache he spun his line overhead and let it fly. Out in the muffled and volatile Pacific a yellow shark heard the bait hit the surface of the deep. Around Joaquín's insignificant circle of light the world was dark, and he stood mythic at the edge.

The duende of that image will never leave me—darkness, spontaneous creation, ancient culture, blood, mysterious power, irrationality, earthiness, the possibility of death.

Finally, inevitably, the very last morning arrived. We said nothing. Together we made the bed and gathered our things. As I brushed my teeth in the kitchen sink I watched Joaquín. He was picking up partial boxes of matches, two from the eating table and one from beside the bed, two more from the kitchen. Now he stood by a window with the matchboxes, consolidating them, a strange and meaningless task for a man so concentrated and potent. His dark face was carved with something I couldn't quite read, a complexity of sadness and longing, like a riptide beneath an inscrutable surface. I couldn't stay with the splendid, sphinxlike sensualist. I couldn't stay, he couldn't leave. What I could do was remind him of what he did not have because he kept tamping down his heart, blowing out its flames, hiding a great and consuming love behind his manifold edicts and blusters.

He looked up suddenly. A precise shadow descended across his face. He put down the matchboxes, turned away, and went out the door.

By noon we reached headquarters, returning by beach rather than by forest. I exchanged addresses with María and strapped my rubber boots to my backpack. When the time came to leave, I couldn't find Joaquín. I departed Cabo Blanco by motorboat with one of the guards, Lupe, who would deliver me to Montezuma.

In Montezuma I caught a taxi to Puntarenas. It was an open-sided truck, and I was riding on the back, talking idly to another passenger, when a motorbike caught up with the taxi. Its driver wasn't wearing a helmet, and I could see clearly the driver was Joaquín. I gave him a big wave and a smile.

He kept trailing the taxi. Perhaps I'd left something behind.

The taxi stopped to take on more passengers. Joaquín stopped too. I hopped down to ask him what was wrong. He switched off the bike and said that he had decided to go to San José. He had business there.

"I was hoping to see you before I left," I said.

"Me too."

"I wish I could go with you," I said. "But my backpack will never fit on the bike."

He smiled. His black-coral totem clacked against the crocodile tusk. "No," he said. "And you have other places to be."

The taxi driver appeared to be collecting fares. We had only a minute. "I thank you for everything," I said. "For all the things you showed me, for sharing your place with me, for caring about nature, for saving me from drowning."

"It was easy," he said.

Two tears slid from my eyes. "I'm sorry to leave," I said. "I have been changed permanently. I will never forget you. Or Cabo Blanco."

"I'll never forget you, mi loca," he said.

I returned to my little boy and resumed mothering in my own eco-system. I received a letter from María some months later. She sent greetings, said it was summer and San Miguel was always beautiful in summer. I wrote her back that I was still trying to write poetry, grinding words into powder, a literature of dissolution and distance, singing toward the tops of trees like the commonest bird. She said that Joaquín was doing well, he had not changed.

Years passed before I tried to find him. I sent a letter to Cabo Blanco and heard nothing. Another year passed. I wrote the National Park Service. A man kind enough to reply said he was sorry to report that Joaquín was dead. He had been transferred to another park, one even more remote than Cabo Blanco. The employee said the new place, a far island, had suited Joaquín, with his mysticism.

Those were the man's words. (The place was Cocos Island, surrounded by coral reefs and known for a stunning biodiversity, now a World Heritage Site, situated 300 miles off mainland Costa Rica.)

Joaquín had married and had a little girl. One day he ran out of insulin, a mistake that proved fatal. He slipped into a diabetic coma and died around midnight, seven years after my sojourns took me to Costa Rica and brought me home again. Joaquín's obituary said he was forty-three and left behind a two-year-old.

A great irony was that duende couldn't keep Joaquín alive. In fact, duende took his life.

Since that sabbatical in Costa Rica I've seen many beautiful, pagan places, some of them coastal, some montane, some inland plains, some piedmont, some marshy, some snowy. I've never seen anything like the endless beach at San Miguel, black sand strewn with bones and driftwood, a thumping ocean playing the rocks like an udu, the blues and yellows of fish flashing, and a heartbroken man standing within a tumble of edges casting to sea, trying to strike one clarifying and unforgettable match.

I think about how the ocean is rising to cover the wild paradise of San Miguel. The thought makes me mute, although the muteness is not, of course, without noise.

Joaquín has entered the great silence that is not a silence, but I still see him occasionally. I see him in dark skin, in bear-claw and eagle-beak necklaces, in long black hair held in place by a leather thong. I know him when I see him, and I briefly relive the moment I watched from the back of a truck as he gave a little wave, inadequate and helpless, and he turned left while I went right, fading into a dark movement of green.

Bird-Men of Belize

ABOUT SPLENDOR

Belize, Central America

The truth is that we have only begun to explore life on earth.
—E. O. *Wilson*, The Future of Life

Soon after we arrive in Belize City on a clear and sunny day in early March, someone comes running to announce that a palm tree in the hotel courtyard is full of scissor-tailed flycatchers. I grab binoculars, eager to see every flying, swimming, running, crawling, bobbing kind of life there is. A scissor-tailed flycatcher sounds marvelous.

Delicate ivory-and-peach-colored birds are bending gracefully to palm fruit. Maybe the head is more of a pale gray than ivory. One of the flycatchers lifts into the air to approach the tree from another direction, and when it flies, its long, stylish tail forks into two narrow and elegant fans. The bird is at least fifteen inches long, and I'm amazed the tail isn't more of a hindrance. Later I will see these birds in the lower Great Plains of the United States, but in this moment I've never imagined such exquisite beauty.

I am in Belize for ten days with a boisterous group of travelers from the Environmental Leadership Center of Warren Wilson College. The trip was imagined by tall, svelte John Huie—then the inspired director—and led by Paul Bartels, a handsome, outdoorsy, and smart biology professor. Our mission is to see biota. Although Belize is tiny, slightly smaller than Massachusetts, it is home to an astonishing 540 species of birds, at least. It claims 4,000 kinds of flowering plants, 700 of trees, 155 of mammals, and about 140 of reptiles and amphibians.

Like most places, Belize has been logged, except in its most remote sections. Because logs were floated coastward along waterways and because industry often sought a particular tree such as mahogany, some of the interior of Belize was saved from the clear-cutting (and thus the road-building that would have brought development). Despite deforestation, Belize has been able to brand itself an eco-destination bar none. More than half the plant and animal species of the world are thought to occur in tropical rainforests, and Belize is a hotspot. Plus, off the coast sits the great Belize Barrier Reef, home to more than a hundred species of hard and soft corals and hundreds of species of fish.

I start a catalog of biota: *Hotel Grounds, Belize City. 1. scissor-tailed flycatcher, 2. great-tailed grackle, 3. northern jacana, 4. great kiskadee, 5. hooded oriole, 6. rock dove.*

LAMANAI

Two mornings later I leave a simple wooden cabana amid foreign birdcall for a hike through the jungle of Lamanai, a Mayan ceremonial center in the north of the country.

This morning four of us have hired a bird guide, sight unseen. We have been hearing about Belize's bird guides. The bird-men of Belize are famous—legendary, actually—trained for their work as for any job and licensed. A rumor has circulated among our group that Belize's ministry of nature is also its ministry of tourism, which would be amazing if it's true.

"Your guide will be Rubin," the woman at the lodge had said.

"Is he good?"

"Rubin knows birds," she said.

Soon Rubin, well-dressed in outdoor gear and freshly shaven, his hair impeccably combed, arrives to lead us off into the darkness. He can't be more than twenty-one—too young for a naturalist, I think. When I was his age I could name hardly ten birds. Rubin is poised, friendly, conversant in perfect English. He was born across the line in Guatemala.

"Ready?" he says, wasting no time although sunrise is a half hour away. Within a few strides on a rough, machete-cleared trail leading toward the Temple of the Jaguar, Rubin spots a northern waterthrush hopping ahead on the ground. All I see is a blue-gray blur as the warbler flips under a disappearing coverlet of night.

"Did you see it?" whispers Doug, a jokester somewhere in his sixties.

I shake my head, grimacing. I've got to get faster with the focus.

"You can see it back home," Doug says. "It'll be leaving soon." Early March, the migrating songbirds have not yet left their wintering grounds.

"Not good enough," I whisper. I want to see everything, now. I'm here to see everything.

Within seconds Rubin cocks his ear toward a croaking and says quietly, "Yellow-billed cacique." I, of course, have never heard of it. Rubin pauses long enough for all of us to focus our binoculars, and this time I get a good look at a blackbird with a startling yellow-green beak and a golden eye. So many of the birds of Belize are not migrants and can never be seen where I live in Georgia.

This is not going to be a lazy stroll. I hitch up my cargo pants and pull out a pad. I start scrawling a record as we march onward. Within a literal minute Rubin has pointed out a red-throated ant tanager.

I have always been in love with the world, as long as I can re-member. Even as a kid I found a peace in nature I could never find among people. When I was in college, someone taught me that *pishing*, making the whispery *pish-pish-pish*, will attract birds, espe-cially in the spring. One evening sitting under a low-hanging laurel oak I called in a yellow-rumped warbler, as I later learned from a field guide, to within a few feet of me. I held my breath, amazed. Somehow that flying ode with its fine cloak of feathers held all the wisdom that could be learned. After that, I wanted to know the birds.

One year at the Florida Folk Festival I noticed a little boy stand-ing near the Cajun jam tent, at the edge of a grove of live oak trees, his sweaty little fist held up in the air. He couldn't have been more than three feet tall, maybe five years old, wearing thick glasses, his face scrunched in hopefulness. Something about the child's face ar-rested me. "Hi," I said to the little boy. In the tent the Cajun fiddlers were jumping.

Without speaking the little boy turned his face fully toward me and opened his hand. In his cupped palm was a half-teaspoon of moist millet and a sunflower seed or two. His eyes were dazzling behind the little glasses. I knelt down. I had no idea where his parents were, but I'm sure they were watching. "Wonderful," I said. "Have any come yet?"

"Not yet," he said. "But one time a bird came."

"I hope lots come," I said.

As the years passed, I knew more and more birds by sight and a few by sound. I was never, however, a natural birder. I never fell asleep with a bird guide open in my lap. Instead, every day I'd have to start over. Which of the vultures, turkey or black, had silver tips on the wings? (Black.) Which had the red head? (Turkey.) How to tell a Carolina chickadee from a black-capped? (More gray on the outer tail feathers of the Carolina.) I've always had to work at birding. I have the heart for it, but not the mind.

And these Belize birds: I could never have dreamed into existence the pale-billed woodpecker, squirrel cuckoo, or blue ground dove. They are endlessly fine: keel-billed toucan, red-lored amazon, black-headed saltator. They get more and more razzle-dazzle: ivory-billed woodcreeper, collared aracari, northern bentbill. They are art, brilliant and alive, their names radiant and tumultuous: chachalaca, violaceous trogan, blue-crowned motmot, bright-rumped attila.

Not only does Rubin recognize the birds by call; he also identifies them by song. Every bird we hear, he names.

"How many birds do you know?" I ask him.

"More than five hundred."

I am incredulous. "How did you learn them?"

"I was trained to be a bird guide in Belize City," he says. "I study hard. I still study them."

"You are an excellent birder," I say. "And so young. In my country, I'd be hard-pressed to find a twenty-one-year-old who knows a quarter the number of birds you know."

Rubin smiles.

Later I research how a person might train to be a bird guide in Belize, and I find very little. The Department of Tourism wants quality control, it says in its master plan, and it names an organization called the Belize Tour Guide Association, which trains and certifies guides, but I cannot find any way to actually apply. I don't know how long the course lasts or how many birds a guide is required to memorize or how much it costs. I find guides that operate without licensing. I can't find Rubin listed anywhere. All I know is his first name and that he is remarkable.

We reach a place where the foot-trail meets a two-path road and we travel along it. Archaeological workers pass, heading to work at the Lamanai ruins. We are within walking distance of the Mayan ceremonial center, where ancient temples are being uncovered by platoons of workers. Some walk and some ride bicycles; occasionally a truck hauls a load of workers. Later I will see them shoveling earth down chutes, then hauling the dirt away in wheelbarrows. From the passenger window of one of the pickups a man speaks to Rubin.

"Pajarero," he says by way of greeting.

Pájaro is "bird." *Caballo* is "horse," and with *-ero* added, it becomes *caballero*, "horseman, cowboy." What the man says to Rubin, then, is "Bird-man." Of the birds.

I look at Rubin. "Bird-man?" I ask.

"Sí," he says. "Somos pajareros."

We are bird-men, bird fanciers. What I didn't know is that in colloquial use the word also means "happy people."

Rubin finds yellow-throated euphonia, plain xenops, black-crowned tityra, and laughing falcon.

"He's going to get a warbler neck," Doug whispers, "looking straight up like that."

"No," Craig whispers back, "he got a special hinge put in." Craig, with glasses and neat silver hair, is a hilarious medical doctor from Florida.

"Like a snake," Paul says. "He can disarticulate his neck."

Like a teacher, Rubin ignores the tomfoolery and keeps moving. He shows us a rubber tree, oozing gum, and launches into a mini-lecture on the gum's significance for birds. Our brains must have reached overload, because at the very same moment, Paul and I both pick off a strip of latex and attach it under our noses.

"This," says Doug, "is the great intellectual moment of the trip."

Army ants sail by with their scraps of leaves, and we skirt a swarm of killer bees living in a tree. Flocks of wild parrots, animals I've only seen in cages until this miraculous moment, fly overhead, like they used to do in the southern US (Carolina parakeets). In the treetops is where parrots belong. And then Rubin spots a toucan, a bird every schoolchild knows from coloring books, perched on a limb, completely at home in the jungle. It's just sitting there.

My mouth is hanging open.

That morning I add six flycatchers to my inventory, including yellow-olive, ochre-bellied, great crested, dusky-capped, boat-billed, and brown-crested. I wish I could describe each one to you, and tell you how we see it, what it is doing, and how it looks, but the movement of the morning is quick, too quick for notes—a word, a gesture, a hurried lift of the binoculars, a snatched breath.

Later I will fidget with my journal under a strangler fig burdened with fruit, at the feet of the sacred Temple of the Jaguar, looking up into branches thick with warblers. Magnolia, Kentucky, prothonotary, yellow-throated, black-throated green. Soon these birds will return to their nesting grounds, my homeland.

I know them better now that I've seen their other country, the other side of them.

NEW RIVER

We see a billboard advertising coffee that says, *You can sleep when you're dead.* We take it as our motto and hire another guide for a spotlight safari on the New River. This thin, darker guide is not a bird-man although he, too, knows the true name of almost everything he sees. The boat goes too fast in the darkness, as fast as a huge meteor that streaks across the glittering sky. On this speedy night voyage we stop occasionally and I am introduced to two kingfishers I didn't know existed: the pygmy, not even six inches tall and very cute, and the green, colored like leaves. In Georgia we have only the belted. We see green-backed heron, white ibis, northern potoo, bare-throated tiger heron. Racing through the darkness, we flash past Morelet's crocodiles, their eyes aglint in the spotlights. Had we been going more slowly, we would have seen hundreds, I think.

The next morning I meet some of my traveling companions on the dock overlooking the river. "Anyone up for an Early Morning Crocodile Snorkel?" Doug asks.

TURNEFFE ATOLL

That morning we leave immediately for the coast, to Turneffe Atoll, called by Charles Darwin in 1842 "the most remarkable reef in the West Indies." By afternoon we're on the water. I have no idea what I'm about to see, what in that moment of ignorance already lies below us.

Our guide, Cristobal, a brown, compact, bright-eyed twenty-four-year-old, ties our boat to a buoy anchored around a coral boulder. There won't be any birds here, I think, but I'm eager to see what's underwater. We splash off and swim up the beds of turtle grass, as directed, so the current will bring us back to the boat. The day is windy. We are in what are called patch reefs, which are small, isolated outcrops. I swim toward large, dark shadows in the turbid water, and soon I'm floating over mounds of coral fifteen feet below. Below me a diver appears, swimming like a fish, effortlessly. It is Cristobal.

The water has cleared some and I see, too, a single track through the sand of the sea bottom. Cristobal swims along the trail and comes to a queen conch. He picks it up from the sea floor and brings it to me. I hold it and nod, breathing through the snorkel, glad he wants to show us things. He returns the conch and heads off through great canyons of coral, motioning for me to follow. I suppose Cristobal has decided to be my personal guide.

The canyons are populated with butterflyfish and angelfish and sergeant majors. With two fingers Cristobal mimics another conch trail, then points toward a lovely parrotfish, its turquoise body dotted with yellow and mottled with orange. I swim behind Cristobal through a coral pass, crossing a pastel ledge that flashes with the sparklers and cherry bombs of tropical fry, and come upon what I instantly recognize as a southern stingray, like an ancient and oceanic heart, twenty feet below in a pool of ash-gray, soupy sand.

This animal can grow to twenty feet in diameter and can be ferocious. We are in its sphere, close, too close. I struggle in the dashing waves to paddle backward, to give the stingray a wide berth, and I see Cristobal, still at my side, cast a long glance at me. He can tell I'm not a strong swimmer. I never even learned to swim until I went off to college, so most of my swimming is academic, unlike someone raised by and in water. Cristobal does something odd then, especially given the secret he tells me later. Or maybe, given his secret, his action is not odd at all.

He takes my hand and holds it, underwater. Below, the stingray lies calmly in the sand, its great ridged tail outstretched, its two eyes watching our two masked figures floating at the surface. The guide is telling me that I'm safe. Beside him, I lie for many minutes suspended on the Gulf, horizontal and superficial, in a borderland between water and air, beholding a stingray the size of a trampoline. Awe tastes like saltwater leaking down my breathing tube.

Then Cristobal moves away to summon other snorkelers. He swims back with them, and after they admire the stingray, he motions us to come along. We kick through side canyons into deeper canyons, colonized by fans and whips and fingers and rods and knobs, boulders and mounds and crusts. Brain coral, fleshy coral, flower coral, leaf coral. Although the names are reminiscent of horticulture, this is not a botanical garden but a zoo, an underwater zoo in which the animals, the corals, are caged to the sea floor, and to each other, and to the calcium carbonate, which corals use to build their awesome skeletons. The waves are the zookeepers, bringing algae and flowering plants to feed the coral animals. I never knew that such a world as these coral reefs existed, populated with peacock flounder, yellow goatfish, bar jack, wrasse, annelid worms, and peppermint shrimp.

I have few words for the feelings I have when I'm under the sea and in the sea, in that luminous beauty, inarticulate, yet language is flowing out of me. Words are made of air. This is not air. The medium for this communication is water. This is not talking. It's not movement. It's not current. It is a current that happens just before movement. It is silent speech. Perhaps I cannot find a name for the slow-motion weightless briny wordlessness because I am not marine and, in the end, cannot speak the language, although I do understand that it is being spoken and can even think I understand a phrase here and there.

An hour later I surface, succumbing to mammalian cold, exhaustion, and shortness of breath. I am loath to reenter humanness and leave the world of stripes and spots and fiery colors, a world of shape and form without shape and form. The underworld is a kaleidoscope of painted glass, it is a silver mirror catching the dancing light, it is an aboriginal festival. The underworld is not hell, but an unabridged paradise found.

When I heave myself against gravity up the ladder into the boat, I start to shake. I bury my face in my towel, so no one can see me, and I weep. There is so much crazy beauty.

HALF MOON CAYE

The next day we go to Blue Hole, an immense deep-blue circle made famous by Jacques Cousteau. The sea has five- to ten-foot waves and the boat is crashing through them, a constant *blam, blam*. "I said 'Sit in the back' and no one listened," Cristobal says to me, rueful. He's learned that I speak Spanish, and he assumes that no one else understands what he's saying. "They have hard ears." Everybody is in high spirits to be in the citrusy Belize sunshine, and nobody minds being slammed around. We all hang on and watch the flying fish, which sometimes sail for fifty feet alongside the boat.

I ask Cristobal if he trained as a naturalist guide, as a bird-man. No, he taught himself, he says. He got a job at the resort and liked it so much that he began to study field guides. He wants to take the course in Belize City, but doing so isn't feasible.

At Blue Hole, an underwater sinkhole, a perfect blue-green circle 1,000 feet in diameter and 400 feet deep, divers go down. I have not yet completed scuba training, so I stay on top, but even if I could dive, I don't think I'd descend. The dive is as much a shark-watching trip as anything. The dive captains load up with chum to feed the sharks, which seems insane because it is.

Directly underneath the dive boat, for example, is a barracuda about six feet long. Eyeing it warily, I move out, snorkeling above the rim of coral around Blue Hole. Trunkfish, blue tang, rosy razorfish, goldspot goby, reef shark. Yellowtail snapper, trumpet fish, damselfish, sea cucumber, spiny lobster.

The sun is burning as divers surface and snorkelers return.

"Taking two aspirin will prevent sunburn," Craig says. He's the doctor.

Dan, a talkative birder, rejoins, "Taking one aspirin will prevent pregnancy." Everybody laughs. "But it's true," Dan says. "You hold it between your knees."

There's more laughter.

"It's true about the sunburn," Craig says.

The divers report that the dive was 139 feet deep, twenty-three minutes. The divers saw four sharks over eight feet long, including blacktips, which can become aggressive around food and have been known to attack humans. The stalactites, large drippings of porous rock, were unbelievable, Paul says.

The boat roars on to Half Moon Caye, Belize's first natural monument, a breeding preserve for endangered red-footed boobies, white-feathered and slate-beaked, and huge magnificent frigatebirds, brownish-black with sharply divided tails. It's March, so reproduction is in full swing. Numbering in the thousands, the boobies nest in ziricote thickets, within which the government has erected platforms for observation at treetop level, affording an astounding view from the middle of the raucous, chaotic nursery. Frigatebird males puff out red throats to attract mates, lightly lifting their long-winged, fragile bodies into the sky. Females hover over the pedestals of their nests.

When it's time to go, I walk back to the boat beside Cristobal, the others trailing. That's when he tells me his story. He is in Belize as a war refugee, he says. Son of a farmer, he was born in El Salvador, in the middle of its civil war, when left-leaning guerrilla groups tried to take down an oppressive government, or at least that's how I understand the horror of it. As a boy Cristobal was rounded up to fight for the revolutionaries, he says. He was forced to kill.

He halts and turns to me on the trail. The others are far behind. He speaks to me as if I'm a priest and he's at confessional, although I suspect that this story has never been offered up within the walls of church. I'm not sure Cristobal has *ever* told it. Surely he suspects that this is the way to forgiveness, that some mercy of Half Moon Caye, where over 120 bird species have been recorded, and some kindness in me can hear the story and understand it and absolve him of his grief.

The first time, he says, his captors made him cut the heart out of a live man.

"That's horrible," I whisper. There is no need to ask why, because in war very little makes sense. War itself is senseless, a compilation of all its violences and atrocities, this one included. But I ask it anyway: "Why?"

"If I hadn't done it, they would have killed me," he says. His face contorts.

"But why?"

Cristobal shrugs. This is a question that keeps him awake at night. He says he was young, he was scared. This was training of a kind. The soldiers had tied the man to a tree and put a knife in Cristobal's hand. The man was begging for his life. Finally Cristobal can't go on. I hear the voices of our friends, getting closer.

"Oh, my god," I say. "I am so sorry. I am so, so sorry."

"There were others," he says.

"I'm sorry."

Cristobal is going to cry, and when he begins to cry, he will sob, and when he begins to sob, he will wail, and then he will be unable to stop. I will take this young man into my arms as if I am the mother of the world, and I will hold him as he wails.

He glances behind us where the first clot of our journeymates have rounded a curve of the trail. "It's okay," he says.

"It's not okay," I say. "I'm sorry."

When I step into the boat, I want to leave that story behind, but of course it rides with me. The story will never leave me, not ever. Even if I get dementia, the story will be lodged inside me.

I feel my heart beating inside my chest, underneath my left shoulder blade, behind my left breast. It skips a beat. It shudders a little and takes up its job again. So we return to Turneffe Atoll, where the wind never stops blowing. The next morning, our day of departure, the sun is a colossal orange ball rising quickly out of the east. Paul, the biologist, strolls down the beach as he waits for our group to congregate. He spots a Portuguese man-of-war stranded by the wind in shallow water. It has trapped a small fish in its tentacles. Two black-and-white fish swimming nearby are immune to the man-of-war's toxin, Paul tells me. There is so much to marvel at, so much to know and see, amid so much toxicity and poison.

I know my travelmates wondered why I hugged the guide so long and so hard when it came time to say goodbye.

JUNGLE

Chan Chich Lodge in northwestern Belize will be the last stop on our tour of paradise. Headed there by bus, deep in rainforest, we ask our driver to stop so we can identify a bird sitting on top of a snag. It turns out to be a knob of wood.

"A green-tailed stationary treetopper," someone calls.

There is more joking now. Everyone is excited to be heading to the jungle, into the possibility of seeing a jaguar.

"Well, if we don't see a wild one, maybe at least they'll show us one in a cage," Doug says.

"Could I take a picture and pretend it was wild?" Craig says.

"You could get in the cage with it and I'll take your picture," Doug retorts.

Dan joins in. "You could probably ride it if you really had the guts."

"If I rode it, you'd be seeing my guts," says Craig.

At Chan Chich we don't see a jaguar. But the air at Chan Chich is filled with jungle sounds, more languages I can't speak. Packs of howler and spider monkeys move through the trees, filling the canopy with a chattering rambunctiousness. Chachalacas, chicken-like, fill the sunrise with eerie and heartbreakingly lovely calls. The males gather in packs to chorus every morning. "They're telling lies about last night," Doug says. I am beside myself with all of it, with white-collared peccary, kinkajou, gray fox, yellow-bellied gecko.

Hiking the wild trails, we add forty-three new birds: white collared mannequin, oscillated turkey, melodious blackbird, crested guan, bat falcon, plumbeous kite, rufous-tailed jacamar. Near the lodge a few trees are filled with the nests of Montezuma oropendulas, which I sit and watch for hours. The fascinating blackbirds weave long, teardrop-shaped basket nests that hang like six-foot ornaments in colonies on trees. They are large birds, the males up to twenty inches (the females smaller), and brightly colored, with chestnut and black bodies, blue patches on their cheeks, wattles pink as peppermints, and large yellow-gold tails. They are a raucous lot, their various calls described as water pouring from a bottle, branches breaking, and whips slashing.

We see swallow-tailed kites, gorgeous black-and-white raptors with deeply forked tails, marvelous aerialists, a flock of them that I greet like the old friends they are. They have begun their slow return to North America, no doubt, working their way up the landmass. In another month I may see these very birds back home.

Then we are saying goodbye to the jungle, and after another night in Belize City, we bid adiós to the country that loves people who love wildness. Belize begins and ends with a register: sulphur-rumped flycatcher, long-tailed hermit, red-billed pigeon, mottled owl. Yes, each of these is a name written on paper, but in my memory and in my knowing, each listing is more than the poetry of its name. It is an alive and very beautiful animal, as much in love with and at home on this planet as I, its every action a dedication to spooling out a simple and fulfilling future.

Snapshots of a Dark Angel

Three girls about ten years old round the jutting flank of slickrock where I sit. They have come running out of a labyrinth of canyons in Arches National Park.

"Can you tell us what time it is?" the dark-haired one asks breathlessly.

"Four seventeen," I tell them. "But I'm from another time zone and I'm not sure if it's three seventeen, four seventeen, or five seventeen." I'm not much help.

The girls have perched on a rock below me, sneaking a rest. "We're lost," the chubbier girl says. "We went playing in the canyons and we can't find our way back." Their faces are not yet desperate in the way older children get.

"Where are you supposed to be?"

"Devil's Garden," one says.

I tell the girls to follow the trail to the road and turn left. I point out a thread of faintly packed trail. "I'll watch you," I say, "to make sure."

They start running one after the other in their shorts and T-shirts, yellow and red and blue. They look back, first one then another, to make sure I'm watching. I wave. The small girl with blond braids who led them here leads them out. She is the one who didn't speak. If they keep moving they'll make it back where they need to be.

I'm in the desert waiting on sunset. Below a wind-carved arch, surrounded by geology, I wait. My skin has taken on the peachy brown colors of this country, legs dusted to chocolate, as if I'm turning to stone. My boy, Silas, rolls his big-wheel truck toward me, making motor noises with his mouth. He carefully points his truck downhill and lets go. The toy gains momentum, rides on two wheels, then crashes into a crumbled rock ledge, beside a scrap of driftwood. Silas goes for it. He's seven, a good climber, comfortable on rock.

We've been camped for three days with our friend Mick, who is silhouetted on rimrock across a juniper and sagebrush flat. Mick kneels to examine something, probably the silvery blue and mustardy lichen that grow in spirals and splotches. Then he lies down. The melody of his pennywhistle floats into the canyon through a coral-colored arch.

Tourists, two women and a man, appear on the trail, speaking German. Another man passes, binoculars around his neck.

Soon the sun begins to throw small fires. It paints fiery layers, going redder, finally searing half the sky as if it has decided to burn its own house. Watching it, I think how many centuries humans have tried to replicate the colors the sun makes at the cusp of night. To infuse paint with the brilliant light emerging from 93 million miles of dark space is impossible, since all our color can do is reflect. I'll admit, however, that Thomas Moran and Albert Bierstadt came close.

As evening deepens, an eerie stillness that only the desert can manifest falls upon the land. There is a palatial quiet, a ravishing silence.

I believed for a long time that ecotourism might save wild places by monetizing them. Apparently we are stuck with the capitalist system, and ecotourism allows nature to compete in the market economy. We can keep wild areas and also make livelihoods from them.

The next morning we make pancakes and set out through Devil's Garden, passing Landscape Arch and Wall Arch. Two red-tails spiral above a swath of wildflowers. We climb the side of Double O, an arch above an arch, now able to look down on the cravens, Mick's name for indistinguishable crows and ravens, who keep up their squabbles. To the east, Dark Angel, a coffee-brown spire, watches us. The earth around her is so deep-red that I collect a tablespoon. The relentless beauty and the beautiful relentlessness of the land leaves me weak-kneed.

There is a family, two little boys. "Don't slide on your pants," the mother commands. As the day wears on, dozens more sightseers appear. Returning from Dark Angel, a group of older teenage boys clots the trail. They look like fraternity brothers. One has caught a sagebrush lizard and holds it aloft.

"Take my picture," he says to his friend.

"Close to your face," says the friend. "This is a super-zoom."

"Hey, man, that's the poisonous kind," another says. The guy holds the lizard closer, then dangles it above his open mouth. The boys laugh. When the picture is taken the boy with the lizard flicks it, too hard, into a scrubby bush. The boys laugh again and trample on up the hill.

"Why did he have to do that?" Mick says, shaking his head.

"Hey," I call after the boys, trying not to be angry. "You all need to treat wild animals with more respect." A couple at the rear of the group turn around and I repeat myself. "Pass the word up," I say, and one of them yells ahead, "Hey, you all need to treat wild animals with more respect."

"I think it landed in this bush," says Mick. We comb the inch-long leaves of a creosote looking for the lizard and can't find it, although Silas spots a baby lizard underneath the bush, in the sand. Then I find one that looks as if it has been dead for weeks, dangling from a miniature crotch. We don't know where the flicked lizard went. I come back to the dead lizard, thinking to show it to Silas, but when I touch the carcass gingerly, the head moves. I yelp. This dying, almost dead, lizard is the one the boy flung. Sometimes life just evaporates.

I sink in the middle of the trail, hang my head in my hands, and burst into tears. I have no idea why I cry so much over spoiled nature, and not just the big things, like political losses or finding out a beautiful forest is gone. I cry if a single tree falls. I feel the tree speaking through my body. I feel actual pain. The image of a crucified lizard, its tiny eye winking in the brokenness of its body, sticks in my mind. I cry as if tears are all I have to speak with. How can we be so careless with life?

"More tourists are coming," Mick says.

I wipe my eyes and Silas presses close. "I don't like people like that," he says.

I believed that if more people saw nature, they would love it and protect it.

The next day is Easter Sunday. We boil half a dozen eggs and color them with Silas's crayons while they're hot. Three generations of a family, occupying two campers and a tent, are camped next door. The grandmother has an extra chocolate bunny she wants to give Silas. We hide it and the eggs before he wakes; the Easter bunny, in this case a desert cottontail, makes him a happy boy. When we go for an early-morning walk through the campground, hard-boiled eggs are hidden all over the place. A crow makes off with a whole chicken egg in its beak.

Later we push on to Canyonlands, stopping at an overlook that is 6,000 feet high, a cliff that drops sharply for 600 feet, bevels into a scree slope, then plunges thousands of feet to the canyon of the Green and Colorado Rivers.

Walking a short trail along the abyss we pass a young globe-trotter wearing headphones, singing along: *Nothing is real.* In an area where people have been doing that sort of thing, we build a cairn. Somehow we strike up a conversation with a couple from DC. Although he's an Episcopal minister, she talks about philanthropy conventions they attend. We tell them we're students, one studying economic sustainability, one nature writing, one the solar system. Pretty soon the woman drops onto her belly beside Silas, who is drawing horses and octopi with creamy sandstone. She takes up a piece of sandstone and begins to sketch flowers. The adults talk about the economic value of wilderness, and then the woman stands up and they talk about how they decide to spend their money. When we move to leave, the minister says impulsively, taking my hand, "Let's pray." I'm not a faith-based person—I think religion has ruined a lot of God's creation and oppressed a lot of people—but I'm all for spirituality. So we make a circle in the slickrock, rich and poor alike, and the minister thanks Father/Mother God for wild places, for the vigor these places give us, and asks for strength in protecting them. The minister raises his hands at the "amen," forcing all our hands in the air.

We're all traveling together, just visiting the planet.

I became extremely conflicted about ecotourism. The more I studied economics, the more convinced I was that a global extraction economy is killing us and the things we love. A local economy would solve some of those problems, and a sound local economy would not be dependent on outside revenues. For all its benefits, then, ecotourism only functions inside a global economy.

Ecotourism is not an earth-friendly industry. It is simply an industry, often resulting in the Disneyfication of the wild. It comes at a price. Public access is not an inalienable right. Ecotourism without education is a product, and education without transformation is pointless.

One danger of ecotourism is that people go into nature for the same reasons they switch on the television, which is for entertainment. Nature's job is not to entertain. Nature's job is to provide essentials, like clean air, filtered water, mediation of the climate, and so forth, without having to be bothered with us. Nature's job is to keep us sane. Nature's job is to blow our minds. Its job is to provide services that cannot be measured in money: sanctuary, peace, beauty, knowledge, wisdom.

Some things ought to exist outside capitalism, and wildness should be one of them.

We journey on through the red rock desert of Four Corners, the place where four states meet, trying to avoid crowds at Capitol Reef and the Grand Canyon. Too many people are trampling along, eroding trails and crushing life along the trails, fragmenting intact ecosystems, and stretching the carrying capacity of some of these places, while many other places go unvisited. We see Grand Staircase Escalante, Vermilion Cliffs, Bryce Canyon. Our bird list grows: great blue heron, bald eagle, golden eagle, magpie, western meadowlark, kestrel, northern harrier, red-winged blackbird.

I think nature doesn't mind humans. It loves us humans. It waits for the quiet ones, to offer up its visions.

Comb Ridge is a monoclinal wall, formed by a crease in the earth's crust, that divides Arizona and Utah for a hundred miles, like a great orange schism through the furnace of the desert. (Parts of the ridge are protected as Bears Ears National Monument.) We leave the nearest town, Blanding, and drive thirty-five miles on pavement and then twelve miles on a dirt track to hike Cedar Mesa, toward the confluence of Fish and Owl Creeks, sucking down water from gallon jugs. I've been reading Willa Cather's *Death Comes for the Archbishop*, gorgeous writing about the Desert Southwest, and also Ed Abbey's *Desert Solitaire*. Mick is looking for Anasazi cliff dwellings in the canyons, and soon he finds some, ancient homes built on ledges under overhanging rocks: Navajo sandstone for floor, roof, and walls—truly a vernacular architecture. Small steps carved in the rocks lead up to the dwellings. Inside, the rooms are small, made to fit the Anasazi. Flat rocks outline windows and doors that are half the size of modern ones. Sparrow-sized rocks nest perfectly in clay mortar between eagle-sized rocks.

Often outside walls of the dwellings are marked with paintings of hands, made by filling the mouth with white clay and blowing, palm to wall, leaving 700-year-old handprints preserved in the aridity of the climate. Above one dwelling all the hands are lefts; above another each individual has stenciled both hands. Some belong to children.

We find a small granary and climb up. It's deep in rat shit, husks of old seeds, broken shards, and cobs, as if the Native people abandoned it last month. Outside, previous visitors have left shards and flint chips displayed on flat rocks.

Soon, as we hike, I'm carrying Silas's pack as well as my own, and after a few miles he tires. We rest in the shade of a canyon while Mick scouts. It's unbearably hot. Sage, blistered by the sun, releases essential oils into the ovenlike air around us, an acrid, invisible, cleansing cloud. A big black fly comes buzzing through the still desert. The fly lowers, whining loudly as it hovers beneath a shrubby sage, and suddenly three sagebrush lizards pour from the rocks. One of them, fat from its hunting prowess, leaps off a slight incline of cryptobiotic soil, the kind you don't walk on because it takes hundreds of years to grow, and catches the fly, somersaulting as it spins around, loses its grip, and lands. The fly is stunned, though, and quick as a wink the lizard recovers and nabs it.

We set up camp under a stand of flowering Fremont's cottonwood, its new buds delicately green. Its long, drooping catkins will turn dry and cottony, but for now bees forage nectar, until the tree becomes a green and buzzing cloud. Fish Creek runs fifty feet away, and when we go down to swim where the beavers have dammed the creek, the bees are hitting the surface of water, scooping up liquid with their mouthparts. We want to avoid giardia from beaver shit, so we're careful not to get creek water in our mouths. Later we'll collect the water for drinking, and we'll have to add iodine to it. Honestly, the foul, chemical taste of that water is the worst part of backpacking.

Evenings are my favorite. After we eat tuna sandwiches, we climb a nearby rock wall, first to look at an Anasazi granary, then to get to the top. Every evening in the desert we've climbed rocks to watch the sunset. Silas is a goat, nimble across crevasses, skipping along ledges above deep canyons. Red paintbrush is in bloom, as is claret cup cactus.

Equinox has passed and dusk is long. Just before darkness Mick crawls in his sleeping bag while Silas and I explore a white sandstone sill a hundred feet above camp. Avoiding prickly pear, we pick dry tumbleweed and spin it to hear a sharp high *whoosh* as it flies. As we pick our way back into the flat valley of the cottonwoods, bats come out. They begin to dive toward us, jerking upward and away at the last minute to avoid crashing, only to circle and lunge

again. Noises frighten them, as does arm-waving. We stand still in
the soft, calf-high grass of our little oasis, my son and I, watching
the animals come so close to our heads, whizzing in from unexpected directions, that we can hear the staccato *tat-tat-tat-flap-flap*
of their supple wings.

A couple of mornings later we rouse at five to an advancing pha-
lanx of clouds. Mick says we have to get out, that flash floods are
a real threat and roads are impassable when wet. We pack up and
hike, me carrying most of Silas's pack, drive the twelve miles back
to the highway, then the road to Blanding, which is mostly empty,
except for hawks hunting from electric lines. "If it weren't for telephone poles, where would the harriers perch?" Silas asks.

People who love nature in natural places, especially famous nature,
like Yellowstone or the Grand Canyon, don't necessarily decode
that into their own lives. They might build a new home instead of
remodeling an older one and build that home in a beautiful place,
cutting down trees and destroying animal habitat and shrinking
species diversity. Wildness as a destination doesn't necessarily
translate into wildness as a life.

Bryce Canyon is a favorite: a two-mile walk down toward the
Queen's Garden, which features a hoodoo (a pillar of weathered
rock) facing a semicircle of hoodoos. We hike slowly, watch a Steller's jay, and take photos. We climb a little slope to eat a tiny picnic
of chicken breast grilled on last night's fire, cheese, and dried pineapple. I collect a small bag of dirt, purplish pink.

At Coral Dunes State Park we treat the sienna sand like snow,
hiking up a dune and sliding down on our bellies. Racing downhill,
getting sand in hair, clothes, and shoes, is a kind of dry-sledding.
From this vantage the world appears to be crumbling apart but not

going to trash and rubble. It's transmogrifying into something just as useful and beautiful as arches and buttes and canyons, something that transports us backward in geologic time and backward toward our earth-home.

This is why we travel to wild places, to go back in time, back to the earth. The dark angels watch us. They know how we behave.

Las Monarcas

WHEN A NARRATIVE FALLS APART
Angangueo, Michoacán, Mexico

I don't know why we hadn't heard about the extreme weather that had hit Mexico. Maybe we'd been too busy planning our trip. Ignorant, then, we arrived to Angangueo in the state of Michoacán at nine one February evening, tired and hungry from a flight across the Gulf from Florida. I negotiated beyond my husband's patience and secured a room, 250 pesos, about thirty dollars. Most of my life I've been broke, and I've had to travel like a local.

The hotel had tall ceilings, wide hallways painted the color of mango flesh, and a bed that sank in the middle. Its sheets crept off at the corners. For supper, instead of braving the dim streets again, though they juddered late with horns and an officer's whistle, I pulled a mango from my pack, and Raven and I ate, comparing the fruit with the walls.

We recollected what we'd seen on the four-hour bus ride from Mexico City—small, brick houses the same gray as the rock and dirt, gold corn drying in piles on red roofs, altars to the Virgin Mary, Indian women in embroidered aprons with shawl-covered heads, a lake of sickly brown water.

Morning was a night away. Morning was monarchs, millions

of butterflies in one glorious, madcap atrium. Eight kilometers outside Angangueo, we would reach Sierra del Campanario, or, literally translated, Belltower Mountain. It was a sanctuary of 350 hectares where, supposedly, two-thirds of eastern North America's migratory butterflies wintered, high atop the 3,000-meter mountain.

The sight of millions of butterflies was to be my birthday gift. I prayed not to lose my vision and not to lose my legs before dawn.

The next morning we set out early, before the sun touched the gray spire of the cathedral on the square. Roosters were crowing. The minute we hit the sidewalk, someone said "Arriba?" in a high-pitched voice. "How much?" I asked. "350," he said. I shook my head and threw him a smile.

The grimy main street, I saw now, was lined with trucks painted brightly with monarchs and the words "Transporte al Santuario." "Arriba?" another driver asked. Down the line, every driver charged the same, 350 pesos, close to forty dollars for a twenty-minute ride.

I kept shaking my head. "There has to be a bus," I said to Raven. His dark beard was neatly trimmed, and his green jacket made his eyes even more yellow-green than they are. We found a bus stop.

"Gone," an adolescent student in a blue-and-khaki uniform said in English. "Five minutes."

"How much does it cost?" I asked him.

"Gone already," he repeated.

"The next one." I made an advancing motion with my forearm.

"Thirty pesos."

"Let's wait," I said to Raven. We stood on the corner, the road leading upward toward a place that people from around the world came to visit, waiting for a way up.

A wrinkled, leathery man sat on a bench, closed like a book. He opened and said good morning to us in Spanish. "Good morning," I said back, and we fell into conversation, but in a moment he blurted, "I have a friend," and jumped up with surprising agility to flag down a beat-up Toyota. The driver was a schoolteacher, a tall, muscled fellow named Simon who taught school in El Rosario, up near the sanctuary. Curly-haired and wearing John Lennon glasses, Simon was a kind soul in a hurry to get to work. We crammed into Simon's car.

El Rosario seems like a village but is an *ejido*, an agricultural commune similar to the Scottish croft or Israeli kibbutz. It has houses, fields, and a school, where Simon, quiet at the wheel, planning his day, taught. Soon we had arrived at a low stucco building with its bottom half painted Azorean blue. We thanked Simon and slipped him fifty pesos. He told us he'd take us down if we wanted and he named a time, said we could leave our heavy backpacks in his car. We trusted him.

So far nobody had said a word about any kind of storm having passed through.

The day was chilly, sky veiled with early morning clouds that promised to evaporate. The approach to the sanctuary was lined with food venders. Some had set up tables, covered with oilcloth, and chairs, where visitors could sit and eat al fresco. Women jabbed at tiny wood-fired grills, and the air smelled like onions and roasted corn.

"Café?" asked a round, dark-haired señora in a blue apron. She and two other women who worked with her were surrounded by blue enamel pots.

"Oh, yes!" I said, and soon the woman was handing us two crockery cups, filaments of steam rising from them. The coffee was sweetened and flavored with cinnamon. Sitting with a warm mug in the chilly morning, absorbing the place, was so pleasant that we decided to eat breakfast, blue corn quesadillas with beans, rice, and cheese.

"Con salsa?" asked one of the women.

"Yes, thank you. Do you make the salsa?"

"Yes."

"Do you grow the corn in the tortillas?"

"My husband does."

We took our plates and began to eat. "The salsa is hot," I said ruefully.

The señoras laughed. "Too spicy?"

"Hot enough."

Those were fine moments, the women happy that their food was a sensual challenge to us, the day fresh as a new nickel and coming alive, every minute the foggy blue-gray of early morning lightening, the leaves of the oaks greening, and the reds of salvia and hibiscus deepening. Sunlight punctured the valley now. Back home, early February, the trees would be leafless, the morning frostbitten and frigid.

After a while we handed over our cleaned plates and thanked the women. Very carefully, in halting English, the señora in the blue apron wished us a good day.

"And you as well."

We paid our fifteen pesos to enter the sanctuary. Our compulsory guide was clean-cut and cologned, a buff young man named Rodrigo, newly certified and with an uncontained nervousness. He wanted to get moving and we did too. We'd been reading about this place for a couple of decades, since it had been "found"; we had dreamed of seeing it, and now we were so close.

We had no idea that what we were coming to see was not what we would witness.

The hike toward the refuge was strenuous, steep, tranquil at first although as the morning ripened the numbers of tourists grew. We were breathing heavily soon enough.

Thinking of the bird-men of Belize, I asked Rodrigo if he knew only the story of the monarchs or if he had studied all the birds and trees and flowers. His "yes" seemed an obligatory answer. He stayed ahead of us, climbing steadily up the *cerro*, or hill, toward some spectacle of monarchs we could only imagine, casting glances back, likely as he had been taught to do, making sure the climb wasn't killing us. I kept flashing smiles at him. We were now hiking through a forest of evergreens, the ground rocky and strewn with leaf litter and small plants.

"What kind of trees?" I called to Rodrigo.

"Fir," he said in English.

"What kind of fir?"

"Oyamel."

"Why are the monarchs attracted to this tree?"

Rodrigo gave a polite shrug. He didn't know. But his nervousness was fading.

A bright bird hopped between limbs nearby.

"What kind of bird?" I asked.

"Pájaro rojo," Rodrigo said. Then in English that was limited but proud, he said, "Red bird."

I smiled. When we began to climb again, I said to Raven, low, "Nope. It was summer tanager."

"At least we know," he said.

The monarch has a restless life although it lives only a matter of weeks. In another six weeks the butterflies of Sierra del Campanario would cross the Gulf of Mexico, above flotillas of oil rigs, and land on the Gulf Coast of the United States. They would search for remnant milkweed for their larvae, feeding on whatever

blooms they could find. Historically they lapped nectar from extensive longleaf pine meadows that carpeted the coastal plains and laid
rows of pearly eggs on the undersides of pineland milkweed and
butterfly-weed leaves. Now they had to make do or die out.
Fewer than 10 percent of the eggs a monarch lays will survive
through the metamorphoses, to caterpillar to chrysalis to adult. The
survivor eggs hatch and fly off, living from two to five weeks. Each
generation spins northward with the spring, hopscotching their
lifespans up the continent, all the way to Canada. By fall their offspring's offspring's offspring return thousands of miles south, sucking nectar along the way, pausing at the Gulf before charging across
the water to Mexico, to the high place we have been climbing toward.

Monarchs were in marked decline, their numbers down 80
percent in the eastern US. In Mexico, "coverage" of overwintering
butterflies had dropped from a high of 44 acres in 1996–97 to a
low of 1.66 acres in 2013–14. Their decline has been steady, and on
average now they cover 6 acres. The causes are habitat destruction,
tree-cutting for firewood, winter snaps, loss of wild milkweed, and
glyphosate. Protectors wanted them listed as endangered.

Suddenly, before us, the ground changed. Something had been
dumped—there was a distinct line where a scrap heap began. I
stopped dead in my tracks. The material was papery, mottled gray,
like pieces of printed material, not a trash bag full but dump trucks
of it. The ground was plagued with raspy gray shredding, ahead of
us, far to the left and to the right. The air began to smell like an old
bird nest.

"What is this?" I asked Rodrigo.

"Las monarcas," he said.

Every picture I'd seen—starting in *National Geographic* when
I was a girl, when this site had been "discovered," through all the

internet sites I'd consulted—monarchs clung to trees. Never had they papered the ground.

"These are the butterflies? Are they sleeping?"

"They are dead."

Wings were paper, printed with black words. Dead. Hundreds of thousands of dead monarchs concealed the ground.

"Why are they dead?"

Rodrigo spoke in Spanish and I was able to ascertain that they had died of cold. Twenty days before there had been a killing snow. Over 50 percent of the monarchs at El Rosario had died.

I gazed around as Rodrigo talked. Indeed, some of the flowering shrubs around us showed signs of frostbite, leaves and branches withered and brown. In some places dead butterflies lay a foot deep.

"They are dead," I said to Raven, bewildered.

Ahead, an old man scooped wings off the trail as if butterflies hidden beneath the dead might yet be alive, as if he might bring them the warmth of the sun and save them from the trampling feet of the tourists, who would not care what they destroyed in their haste to see a natural wonder and to say they had seen it and take pictures of themselves with it, to show their friends back home. By saving monarchs the old man might save his people.

"Jesus," I said to Raven.

"Climate change," he said.

"How far up are we?" I asked Rodrigo.

He looked blank.

"How much farther?"

He shrugged and I briefly wondered if he enjoyed his job. I had asked the question because I was trying to get a sense of the extent of the loss. I'd never seen anything remotely resembling this cemetery of monarchs. There were enough dead to scare anybody, unbelievable numbers dead.

Rodrigo started walking again, his boots coming down severely in the mulch of butterfly wings. The old man paused to watch him. Did the bent old man ever weigh the heart of a guide against the heart of a tourist? Did he weigh his own heart against Rodrigo's? The man's heart hung heavy, that much was obvious, as he raked the mountainside with his bare hands.

We had not climbed much farther when Raven paused to point out live butterflies hanging in ragged gray masses from the limbs of oyamel. Trunks were wrapped with monarchs. The morning being chilly, overclouded, the butterflies remained enfolded, wings closed for warmth, clinging to trees and to each other. In the grayness of the cool forests their mood seemed somber, as if only spring flight might repair their arduous loss.

Below the live masses, dead monarchs papered the forest floor even more thickly, a choking humus of powdery, orange-black-and-white leaves. Now the air smelled like the hold of a ship, anchored at port, and also like rotting pumpkins. The butterflies were deep enough to wade through, deep enough to drown in. We began to suspect that more than 50 percent were killed, maybe even 75 percent, because there seemed to be more dead than living.

We thought about what the refuge would have been like had the killing snow not swept in, and, as if we had been cheated, were flooded by a yearning to return, another year, to see the beauty en masse.

But the mood of the refuge was shifting. Where the sunrays touched, living butterflies would begin to open their wings and color the trees like a Halloween party, and as the day ripened, they began to flutter about in dispersing clouds and cyclones and waves. Airborne monarchs moved in weather patterns.

We reached a green plateau, and for a while Rodrigo allowed us to sit and drink in the viewscape. We watched the live butterflies dance and dabble, now thousands of them filling the atmosphere like confetti, like tiny balloons. Humans love bright things. We love anything that floats.

Later Raven and I found answers. Two weeks earlier, between January 12 and 15, heavy rains had fallen, followed by temperatures down to 18 degrees Fahrenheit. This would become known as the historic storm of 2002. Scientists counted over 2,200 butterflies per meter dead. Eighty percent of the butterflies at El Rosario froze, between 200 and 272 million butterflies gone. Just like that.

As the time approached for school to dismiss, we walked back down the sanctuary trail with Rodrigo, tipped him, and said goodbye. We sat in the grass beside the dusty road, looking down at the village of El Rosario. A flycatcher landed on an electric line. A footpath led from the road to the school, past a wooden fence, past two oxen unhitched from their yoke now feeding from a stack of fodder, past simple houses made of material found close at hand, past an iron bell on a pole. The houses were more open than American houses and more colorful, painted in dazzling shades of seagreen and lapis lazuli, bright with red geraniums in pots. Between the place where we sat and the footpath to the school corn grew in a cascade of rocky fields, from new-harvested to fresh-plowed. A horse grazed in a pasture, and somewhere a donkey brayed. Beyond rose a cliff of dirt, into a hill with one side shaved off, and then on into mountains, the closer ones a pine green, the more distant ones fainter, a frosted gray. On a far mountain a satellite tower received signals from all the continents of the world.

A man picked up the oxen's yoke and disappeared. An elderly señora watered pots of pastel impatiens. A woman traversed the footpath and entered the school. Children began to stream out and run toward the homes.

"Their lives are in plain sight," I said.

"They come in contact with each other more than we do," Raven said. "The generations seem to live together."

"It's more of a community."

"They seem to live at a slower pace," Raven said. "More of a sustenance life, not running around."

"A lot happens out of doors," I said. "By hand."

"There's a lot to learn from them."

"I wish my life were more like this."

The next morning we returned to the butterfly sanctuary on the 8:30 bus with a group of students from Dartmouth. The driver looked no more than fourteen or fifteen. He let us off at the *ejido* and we walked to the refuge under a cloudy sky, birding along the way, house finches and yellow grosbeaks and a white-eared hummingbird. Wildflowers I couldn't identify bloomed along the trail and within hedgerows bordering the cornfields. With the same family we ate quesadillas and also fajitas with potato and sausage. We drank the same sugary coffee.

This time we managed to get through the entrance without being assigned a guide, because of the melee there, but that was fine. We knew the trail up and the sanctuary was more wonderful without an escort. We hiked through the devastation without pausing. Soon butterflies hung in freckled honeycombs from the trees, huddled and closed, masses as long as six feet and as wide as two. As the day warmed, the outer insects commenced to unfold and refold their wings, flashing orange, so many tiny flamenco dancers. After a while they let loose from their swarms and flew about in the sun, and then the next layer of monarchs began to blink orange, like little lighthouses.

We reached the flat plain where we'd briefly rested the day before—a treeless llano on the sierra—popular with picnickers.

By 2:30 there was a steady stream of visitors, mostly Mexican. It was Constitution Day, we learned, a national holiday. For a few hours we sat in the sun on the llano, sketching a dead monarch that Raven collected while live monarchs fluttered by in ephemeral constellations. Other daytrippers would see us drawing and watch for a while. One man asked if he could videotape us, I don't know why. His beautiful daughter started speaking to us in very good English. Another man excused himself and asked if he could take our picture.

The best part came later, after we started down the mountain. The forest was sizzling now, the sun overhead. We began to enter a cloud of butterflies. They were waking, opening their wings in huge numbers. Where the sun struck through the canopy, branches of tall firs were set afire with orange. The butterflies dripped fire from twig and stem as they exited their gray masses. The forest filled with fluttering, burning love notes, careening in crazy combustions through the iron-green shade of the conifers.

The monarchs were like brilliant origami. They were like flames released from the prison of a fire. They were gliding candles. I tried to think of other metaphors, but they were so much like fire. Hanging, they looked like ashes, dead. Flying, they glowed. The sun ignited them, until the butterflies became messengers of sun, our only sun. It was as if the sun shattered into little pieces that went soaring here and there, passionately and senselessly.

That evening we ate chicken feet soup in the plaza. For dessert we indulged in a tutti-frutti mélange, something called a Plantain Split, a ripe plantain fried whole and topped with sweet cream, strawberry jelly, half a peach, and condensed milk. Afterward, we set off through the tight, jittery streets in search of *pulque*, a cloudy local drink made from agave, chewed in pieces and spit into urns to ferment. Our search brought us down an alley to a modest house that served as a store where a woman in a long skirt with eyes dark as vanilla extract laughed at our request and dipped two glasses of cloudy liquid from a pot. At sunset, walking, sipping, we came upon a cactus we call Spanish bayonet, tall, with seven trunks, full of house finches. Easily the finches numbered a hundred.

Up on the mountain, as darkness fell, the monarchs drew together in tight principalities, each butterfly a bit of a bigger journey, its life a baton to be passed between Mexico and Canada and all the land between, a brief flare in an incredible inferno.

The Dinner Party

WHILE REMEMBERING RICHARD NELSON, 1941–2019

Sitka, Alaska

Oh, I have made myself a tribe
out of my true affections
and my tribe is scattered!
How shall the heart be reconciled
to its feast of losses?
—*Stanley Kunitz, "The Layers"*

Of a month spent in Alaska, what has remained longest in my mind is a single meal. Everything that happened hinges on it. In my memory that meal has come to represent a whole territory, a place disturbingly rich, not a land of milk and honey and fatted cows but a provenance of superhuman wildness, ten times to the wild power. When the last spoon rested against the last bowl, the rise of agrarianism seemed a pale and pitiful eclipse.

It was April. I'd left my farm in Georgia to spend a month writing in Sitka, a small island city in the filigree of the northern Pacific coast. Food is always prominent in my purse of concerns, and I guessed I'd be eating processed food shipped from far away. Farmers markets would not be operating, with all that snow and ice.

My assumptions were shattered, and in the shattering my thinking about food permanently changed.

I arrived in Sitka on a flawless day with my husband, Raven. The plane circled down and down into a place I'd always dreamed of going. Carolyn Servid, slight with graying hair, retrieved us from the airport and delivered us to a studio apartment at the base of snow-capped Mount Verstovia. Carolyn was director of the Island Institute, a nonprofit that worked at the "nexus of story, place, and community."

I found that nexus. It was at the center of a spiral.

Carolyn handed me the keys to a borrowed car and an envelope containing a $300 stipend, then left us to settle in. She told us to be careful on walks, that grizzly bears were already coming out of hibernation. She said Hank Lentfer wanted to take us out in his boat and he'd be calling.

Out a window we could see Verstovia's rocky peak against the cerulean blue. Out another Thimbleberry Bay lay placid. Beyond it was the channel to Sitka Sound and then the Pacific. An eagle cried and cried.

Hey, air. Hey, waters.

Any passion or profession enlists a person in a tribe. I belong to a tribe of nature writers. Hank Lentfer belonged too. I'd met him at a conference years before and had kept in contact.

When he called, he said he was picking up a new sheet-aluminum boat and wanted to give her a go. Did we want to join him? We could look for whales. Yes? We were to meet him early the next morning at the wharf, downtown by the ANB Hall.

"ANB?"

"Alaska Native Brotherhood," Hank explained. "It's a social club. The Tlingit and Haida use it as a community center."

Sitka sits on the island of Baranof, at the very edge of the sea. Many buildings, including the library, are built on the waterfront. The tide comes in and goes out along a rocky shore, right in the middle of town. When the tide goes out, it leaves in the rock pools of the intertidal zone behind the library all kinds of gloppy and bizarrely beautiful things—mint-green anemones, pinkish common sun stars, bright-orange and purple sea stars. In the other direction, the Pyramid Mountains rise, snow-covered above 2,000 feet.

The sun cast lemon over the wharf as we arrived that first morning. A wrinkled man wearing trousers and sneakers stood in the shallows, eating something he'd fished from the water. He was talking to another man, short and sturdy, and tiny yellow globules flew from his mouth.

At the far end of the dock Hank was waiting with a hug and a handshake. "Yo," he said. "Imagine this!" Hank was tall, with a red-brown explorer's beard and eyes like two dark rockets. He was a man not young, not old, but coming into the fullness of life. He hopped into his boat and turned back with a hand outstretched.

"Maiden voyage," he said.

I looked around. The boat was shiny and clean, but maiden voyages carry some risk. "You have a life vest for me?"

Hank guffawed, his chin sharp. "The water is 43 degrees," he said. "If you hit it you'll die."

"With a vest I'd at least have a chance," I said.

"I'll bring one."

For a hull that could withstand a run-in with a glacier, a hull welded by a master boatbuilder, Hank had paid $12,000, plus a few thousand to add a covered cockpit with a windshield. A new 100-horsepower motor cost $7,000. In this shining beauty we puttered into the sound, past a fleet of fishing vessels. Scores of gulls wheeled and screamed.

Along the shore, impossibly large and striking birds, dozens and dozens, rested on bare limbs of trees. I blinked.

"Am I seeing bald eagles?"

"Those are bald eagles."

Where I lived, in south Georgia, the sight of *one* bald eagle was a phenomenon. Here were five on a single limb, another three on a limb above. That I saw fifteen in one tree was not a dream. "What in the world?" I said.

"It's the herring run," said Hank.

"And that is?"

"It's an ancient event," he explained. "Part of the phenology of April. The herring are laying their eggs and the eagles assemble here to feast on the eggs."

Emotions did not readily occupy Hank's face, but I could identify elation there. He said that herring spawn turns the entire coastline milky white. The roe glues to whatever is in the water, including seaweed, and it attracts birds, sea lions, seals, and whales. People harvest it, right out of the water, to eat.

"That's what the man at the wharf was eating."

"Herring eggs on kelp," Hank said. He said that the tribes gather hundreds of pounds of roe in a narrow window of time. They sink hemlock branches in the sea to collect the eggs or harvest seaweed laden with millions of tiny pearls. They distribute bags to elders and serve them up in community feasts. "This is happy feeding time in southern Alaska," Hank said.

Deep in our history, every human was a forager. In fact, for 90 percent of human history, we foraged. Because they were mobile, hunter-gatherers could not amass capital. Agriculture is what allowed humans to settle down and store food, so the rise of agriculture was the rise of capitalism and I blame of lot of nature's destruction on capitalism. But even the luckiest, strongest, most intuitive foragers were poor compared to agrarian societies, at least in capital— though they were not poor in terms of strength and health, physical ability, and free time.

We can think of human history as a continuum that moves from

us being hunter-gathers to agrarians to industrialists to technol-
ogists—which is to say, from wilderness to farms to factories to
computer centers operated by robots. Modern-day foraging is a
plummet into the ancient, as if I've been sleeping in fields of grain
and now I'm awake.

Nature writing has been called a marginal literature. If culture is
a set of stories we tell about life in a place and how to navigate that
life, then nature writing is literature at its most essential. Its tenets
are that humans are biological; that we are dependent on the earth;
that places are vital to our psyches; and that humans have volumes
to learn from nature.

Back in 1999 I was at *Orion* magazine's Fire & Grit conference
when environmental author Gary Nabhan, in a talk called "Food
and the Politics of Place," said that nature is not simply wild animals, plants, and land. Food too is nature, and much of our food
arrives to us heavily fertilized and pesticided, waxed and gassed and
irradiated, highly processed, genetically modified, from animals
treated with hormones and steroids, and from very long distances.
Such food is hazardous to nature. Nabhan first announced the
idea of "local food," saying if it's grown closer to home it's health-
ier and helps solve the climate crisis. This was years before the
hundred-mile diet became a fad. Gary was eating food he brought
with him from his Arizona domicile, dishes like cactus-pad salsa
and mesquite tortillas. He said he was using food "as a metaphor to
weave my life back together."

If eating is an environmental act, I became an activist in Alaska.
Or perhaps I slid backward along the human timeline until I could
finally see wilderness.

The ocean was exceedingly calm, and water lapped at Hank's boat.
The day was sunny again, stunning, as if at any moment golden

streamers and silver confetti might fall from the blue sky. The weather was unusual, Hank said. Sitka logs 200 days of rain a year. But I didn't yet understand rain, having arrived to a brilliance that had lasted into a second day.

Around us floated an armada of beautiful islands, big and small. Hank said he wanted to take us to St. Lazaria, which I assumed was one of them. The jagged, snow-covered peaks of Baranof Island towered in the distance. Mount Edgecumbe, a volcano on nearby Kruzof Island, was etched sharply against the horizon, a shawl of snow around its shoulders. Edgecumbe's major eruptions 12,000 years ago were thought to be triggered by the retreat of the glaciers, and its last eruptions occurred 5,000 to 6,000 years ago. The Tlingit, who have been in Alaska "since time immemorial," call it *L'ux*, or "Blinking." One April Fool's Day a local jokester named Porky had helicoptered a bunch of tires up into the caldera of the volcano and set them afire. That caused a stir. He got into some trouble for it, I heard.

In front of us something leaped like a disappearing parenthesis. It was black with white sides. "Harbor porpoise," Hank said. Back home, we have bottlenose dolphins, larger than these cousins.

Hank spotted blow-spouts. "Gray whales," he said. By the time I looked, the sea was calm. Hank idled in the general direction of the spouts and cut the engine. Apparently gray whales, mammals up to forty-five feet long, can stay down a couple of minutes between breaths. We sat still. Then before us a head crowned and an immense gray body rose toward the sapphire of sky. Another whale lifted from the ocean and rolled. We were close enough to hear whale breath released in loud pops, followed by an eerie ringing as they drew in air, then more pops. When they blew, fountains shot more than ten feet above the sea in a forked shape, and for a moment two iridescent puffs of rainbow-colored mist, pomegranate and tangerine and azure, drifted down into the roiling ocean. The whales were massive bobbles, their sudden whale clouds kaleido-scoping against the black sand beaches in the distance.

Again the whales keened long intakes of breath and blew sudden mushrooms of rainbow vapor. Then they waved their flukes and disappeared into the bubbling sea.

St. Lazaria is a pile of black volcanic rock no more than a few hun-dred feet across, thought to be the remnants of an ancient, eroded volcano. It is hourglass-shaped, with a high-wire saddle and two ends that rise in large cliffs. In 1974 the feds designated the entire is-land a wilderness, to protect massive nesting colonies of fork-tailed storm petrels, Leach's storm petrels, common murres, thick-billed murres, tufted puffins, and rhinoceros auklets. I didn't know all that then.

The sea was unbelievably calm, but the closer we got to St. La-zaria, the crazier the currents got. Hank eased the boat toward the lee side of the island, nosed her into a rocky cove, and told us to leap off. He retreated and stalled the boat, which had low wide sides and a beautiful wedge, and anchored her in the cove by tying two cleated ropes to armfuls of bull kelp. The kelp moors to the sea bottom by a root called a holdfast, known to be incredibly strong. Still, our craft was tied only to seaweed, amid lashing waves. From shore we watched Hank lift a blue plastic kayak into the water. He was reared in Alaska, his father a bear biologist. He'd been on the water all his life. Thor might have been a better name for him.

Below our feet, St. Lazaria's basalt was smut black and almost fibrous with bubbles, like petrified sponge.

"Pumice," I said to Raven. "I've never seen anything like this." The pumice was splotched with bright-yellow lichens that lived out of the splash. Tidal pools in the basalt were marvelous little gardens

of kelp, algae, anemones, limpets, and sea stars—both the true stars
(up to a foot wide) and sunflower stars (twice as large, with as many
as two dozen rays).

The three of us climbed to the skinny sill. Below, the fierce blue-
gray Pacific hurled itself against basaltic shores, sending spray high.
"I can't imagine the ocean on a *rough* day," I said. In a snag two
bald eagles consulted. We stood listening to gulls and admiring
the handiwork, the intricate carvings, of volcano and sea. A large
pool directly below maintained the water level of ocean through
an igneous corridor where water rushed back and forth, sending
splashes high. Fishing boats—trollers trolling for king salmon with
hook and line, Hank said—studded the horizon, maybe thirty of
them within eyesight. We scanned for blow-spouts. Tiny Sitka had
disappeared and the coast of Alaska was all wildness, all powerful,
all bigger than ourselves.

Our boat floated peaceful and calm.

We didn't stay long. When the time came to go, Hank perched
his kayak at the head of a little rocky canyon. Every couple of minutes a wilder wave would come along and spew our feet, and if we
shoved Hank into the retreat of one of these waves, he wouldn't
have to clamber into a kayak being battered against a rocky shore.
Sure enough, he went out without a hitch in a spume of foam and
soon returned with the boat.

"Do you have a name for her yet?" I asked.

Hank looked sideways in such a way that he seemed to wink.
"We like the name *Bob* for a boat. My wife insists on a girl's name,
so our sailboat was *She-bob*. For this one we're considering *Go-bob*."

"You should think about *Shishka-bob*," Raven laughed.

Slowly Hank looped the island. Melting ice dripped from cliffs.
On the north side a cave opened like a yawn at water's edge, twenty
feet by twenty feet, and Hank eased the boat inside. "When the birds

are nesting here," he said, "the smell is so strong you can't breathe. The noise is tremendous." But this was April. Most of the northern breeders were on their migration home, and the cave was empty. It smelled musty, like old pantyhose. Absurd hot-pink sponges clung to rocks at the waterline, as did more sea stars in neon colors. One flank of the cave was streaked with emerald-green moss. Water dripped all around through cracks in rock. I could barely notice these things, however, in my nervousness, as each wave rocked the boat closer to the jagged rocks. After a minute Hank squeezed out and we went on.

Could I have ever imagined such a place? Could I see myself living with whales? Was I wild enough to enter such wildness, which partners so closely with death? Had I accepted the diminishment of the world with barely a whimper?

On St. Lazaria's ocean side, black rock fractured into a cavern large enough to hold a brigantine. Along its many ledges hundreds of birds, penguin-like, stood. They were murres, which breed in colonies in niches on sea cliffs. Even at our distance we spooked them, and hundreds of murres poured from the cave. Hank cut the engine. A treble of whirring wings resounded off the cliff as the murres took flight.

In a panic humans will stampede and kill each other. But there was no jostling with murres. I don't think any two of them ever touched wingtips. Each leaped off into space and flew out to sea, in formation, then rapidly back, circling into the coastal cave. A few hundred didn't spook but sat watching—how quickly the sea carried us toward the sharp rocks—until Hank hit the engine switch and swifted us away.

Another eagle surveyed from the cliff.

How much life, I thought. *How many hearts. All together, an ocean
of blood and another of sap. A continent of bones.*
I love this world.

If we eat from a place, don't we become the place? Isn't eating a direct way of filling ourselves with a place we love, of making it a part
of us, of becoming it? If we eat wild food, can't we become wild?
Can't wild food get us closer to the heart of wildness?

For the next month I slogged through Alaska. After years of longing, I had found my way to the heart of cold, heartbreak of glaciers,
doggerel of Jack London, scrappiness of the Iditarod, electricity of
northern lights, myth that is not a myth. Eagles shrieked as they
spiraled, and song sparrows recited the longest birdsong in the
world. Skunk cabbage emerged stinking from the thawing ground.
Devil's club drew blood.

I had left spring behind in Georgia and reentered spring, one
of extended twilight and cadence, night skies crowded with stars,
ocean thick with herring eggs, paths bordered with the crimson
thumb-toppers of salmonberry.

While I spent the mornings writing, Raven occupied himself
elsewhere, including volunteering at a salmon hatchery, which kept
60 million fry—king, coho, pink, and chum—in tanks. Someone
gave him five or six frozen whole salmon, that Alaskan staple, and
he came home with them hanging across his shoulders like loaves
of French bread. Alaskans preserve salmon by smoking it and can-
ning it in pint Mason jars, so Raven borrowed a smoker, and soon
we had our own stash of delicious smoked fish, the flesh orange as
the Pyramid Mountains colored by sunrises I watched coming up
over them. The oily gray skin of the salmon shone iridescent.

One of the fish biologists asked Raven if he wanted bear meat—
his family had extra—and Raven did. Forgive me. It was robust and
delicious.

One heavily overcast day in a drizzle Raven and I hiked Mos-
quito Cove Trail past totem poles through a rainforest of old-
growth hemlock and Sitka spruce, green moss like a thick carpet,
ferns lush and verdant. A log bridge had been fashioned from a
fallen tree six feet across, carved flat on top and outfitted with a
handrail.

At times the trail edged onto Halibut Point and we walked the
beach. Herring eggs were sometimes four inches thick on branches,
washed up by the full moon's extreme tides, and murders of ravens
fed. Eagles scooped up great beakfuls of the eggs. Skirting the birds,
we lifted kelp cemented with eggs and ate it looking out at rafts of
buffleheads and Barrow's goldeneyes.

Another day we watched a rehabilitated eagle released back into
the wild. Someone who had worked with the eagle was given the
honor of springing open the door of its cage. For a moment the
iconic bird paused on the rocky beach, then it took off on its wild
flight above the sound and became a soaring black cross in the sky.

We celebrated Easter at the Russian Orthodox Cathedral, St.
Michael's of the Archangel of God, a standing service that involved
a priest kissing gold icons, parishioners kissing pictures of Mary,
and an altar boy kissing the priest's hand as he handed him an incense pot. The choir harmonized as it sang liturgies and chants,
calls and answers, O Lord have mercy. We enjoyed a lunch of Native
fry bread and clam chowder in the church basement and bought
kulich, a yeasted sweetbread baked only at Easter in cans using an
old Russian recipe, with raisins, nuts, and citron.

I judge the richness of a life by its larder, and I judge food by its
proximity to the wild. Alaskans, I found, live closer to their food

than most Americans do. Almost every family plays some part in providing for itself—gathering, growing, drying, smoking, freezing, preserving. Fifty percent of Sitkan families hunt, 75 percent fish, and many of them keep a remarkable supply of fish and game.

If these Alaskans visited our agrarian home in the South, I could show them alligator, but not bear, not bison, not panther, not wolf, not woolly mammoth. My place is no longer deeply wild. To see an alligator we'd ride through miles of industrial farmland, the intent of which seems less about producing food than destroying all vestiges of wildness, including fencerows, ephemeral wetlands, pollinators, wolf trees.

Even considering the wild abundance of coastal Alaska, I was unprepared for the celebratory meal that would be goodbye. It would be a simple meal, Carolyn said when she called with an invitation. But that supper turned into a four-hour feast, the longest in my life, and would come to represent all of my feelings about the entire earth. The meal was a crescendo, an arc of a bridge; and in its shining I could satiate my insatiable cravings for a wilder world.

No, the shape of the story was not an arc. It was a spiral. An airplane had spiraled into Sitka and delivered me to a new landscape and mythos. I circled around and around toward some understanding, as if walking a labyrinth, until I reached a hub at the center, and there at some axis I didn't recognize until much later was a dinner table laden with food, a group of friends tucked in around it.

Although a spiral moves inward, its force is a straight line, under control. A spiral moves outward as well, also with the unremitting force of a line. So a spiral becomes a wheel of fortune, spinning and unspinning, a cable flung into the unknown that vaults out again. I orbit to the center, pass through a revolving doorway, and coil out. An eagle rises in a spiral into the sky in an action called "kettling," riding thermals. Likewise, I kettled into Alaska and kettled out.

Carolyn and her husband Dorik, an older, quiet man, lived off a macadam road out of town, at the end of a quarter-mile trail that wound through a forest of old yellow cedar. We gathered our bags and baskets and started out through the woods. The trail was rough and uneven with roots, but magical, as if we were caught up in a fairy tale and anything could happen in these woods. Soon enough we crossed a shadowy footbridge and the woods opened onto a narrow bench of land beside open water. Carolyn and Dorik's simple, well-designed house resembled a wooden boat anchored at the edge of Thimbleberry. Two walls of windows came to a point overlooking the bay, behind a widow's walk.

An environmental and place-based thinker, Carolyn had visited Sitka a couple of decades earlier and, to her surprise, realized she'd discovered home. She founded the Island Institute to encourage creative thinking about how best to inhabit place. "What if we conducted our human affairs in the Tongass in such a way that the abundance of life was our focus?" she asked. In her memoir *Of Landscape and Longing* she says, "Our commitment begins with the fact of our love." She and Hank Lentfer had recently teamed up to edit *Arctic Refuge: A Circle of Testimony*, an anthology of writers calling for protection of the Arctic.

When we arrived to the dinner party, Hank was already there with his wife Anya, a family practice physician pregnant with their "first and only" child.

Another writer friend had arrived ahead of us too. Richard Nelson, or Nels, was a tall, birch-like cultural anthropologist who had lived with the Athabaskan and Eskimo. He wrote *Make Prayers to the Raven* about his work with the Koyukon, followed by *The Island*

Within and other enthralling books of ethnography and nature. He too lived in Sitka.

The eighth member of the party was his partner at the time, Liz McKenzie, a slender and pretty environmental filmmaker.

Carolyn was setting out smoked salmon with crackers. Dorik was uncorking a bottle of malbec. Hank picked up a sack, lifted from it a foil-covered bowl, and peeled off its cover. The bowl was filled with square green crackers plastered with tiny bone-white pebbles.

"Herring eggs on kelp!" he said. "It's not a party without them."

He set another dish on the table. "Black cod collars and halibut cheeks," he said. He had scavenged this delicacy from a fish-processing plant. Like chicken backs, the heads are thought useless for market, and they get dumped. But their secret pockets of flesh are more flavorful and richer than fillets.

"Dumpster diving," I said.

"Call it foraging," said Hank.

Cheers to Alaska. Cheers to the wild world. Cheers to good friends you may not see again for decades, if ever. Cheers to the herring run, to eagles, to whales, to St. Lazaria. Cheers to passion, to art, to the chance to spend a month in a wild terrain.

Raven and Dorik disappeared into the kitchen, from which emanated aromas of seawater and cinnamon. Raven had responded to an ad in the newspaper selling king crab, available at the harbor at 8 a.m. until it was gone, $6.50 a pound. He had paid $48 for a giant one. Soon the men returned: king crab served with lemon and butter.

Carolyn replenished the wine, and Dorik delivered a platter of fat pink Alaskan shrimp, boiled with lemon, clutching roe into the curves of their bodies. Was this a third course or were we still in the first?

Hank was talking about the "protein-rich land"—how he catches salmon with a gill net, and about twenty-five fish supply them for a year. He and Anya tried a monthlong experiment to eat only from the land. They ate salmon served in every imaginable manner, including with homemade kelp salsa.

"I never get tired of salmon," Carolyn said.

"I watch salmon, fish for salmon, eat salmon, celebrate salmon, and dream of salmon," said Nels. "Alaska should be known first and foremost as the Salmon State."

Everybody wanted to know how the month had gone for the Georgians.

One day we climbed Mount Verstovia, we told them. The trail began behind a bar a couple miles out of town and went very steeply up. The trail was well-built, with log steps crisscrossed with chainsaw hatch marks, gravel in places, and orange, diamond-shaped markers every twenty feet. In some places the trail-builders had sunk stainless steel posts for a cable to help in crossing shanks of pure rock.

We climbed through the dim, mossy woods of old hemlock and Sitka spruce. The ground was a puzzle of fallen trees, tons of moldering deadfall, quick to get bundled in moss but slow to rot.

The vistas got progressively stunning as we climbed, at first the town and near islands visible, then more and more islands, until finally a view far-reaching and encompassing. We now were deep in snow, crusty and crunchy, with some evidence of small avalanches around us, and the wetness began to seep through my leather boots, into ragg wool socks. Marker 185 had "Last Marker" handwritten on it. We gained a knob, a high point, snow all around but the rock dry and warm, and pulled out crackers and brie.

Around us lay a spine of sharp, snow-heavy peaks, the backbone of Baranof, the archipelago, a volcano and its twin, a town with its many streets and rooftops and boats, and, beyond it all, a misty-blue and billowing mass of water, dissolving into the rim of creation.

When I looked out from Verstovia's high peak at the vastness and breadth and power and unapproachability of wildness, I wondered how we paltry and weak humans have been able to limn such a terrible pathological mark on the earth. "I wonder," I said to Raven, "how, with all this, we could have set our foot so harshly against wildness. We have changed the course of planetary history. Here it seems not just improbable, but impossible."

"We will end life for other species," Raven said, "and ultimately for our own, but the earth will go on."

"On an altered trajectory," I said. "Not the one that got us here."

The room was a hive, humming and buzzing, scrambling for the nexus of story, community, and place. These were people who spent time thinking of what the story was, what the place was, what the community was, how the story might save the community or the place, how places might find their stories, how communities might save their stories. These were people thinking about stories of belonging, not stories of separation, and stories of interconnection.

In the best of times clocks tick unnoticed. Hours pass as if they are wind blowing across a bay and around a house. A couple of hours passed before Carolyn announced the main course: fresh salmon, bulgur, green salad, and sourdough bread. It was not an epicurean feast, no fancy preparation or presentation, yet among the memorable meals I have eaten in my life it stands at a zenith.

Here I want to hover above a dinner table at the edge of Thimbleberry Bay. The moon is a day off full and ringed with the palest of green, that circled by a coral-colored halo. The moon illuminates the diners and reflects serenely off the water of the bay. A printed yellow cloth is spread, at the table's center a bouquet of daisies and on the table eight cups filled with wine. Eight people lean forward, reaching toward bowls and platters as they eat the place where they find themselves, and also reaching toward each other.

The diners have teleported backward a thousand years into a primal abundance in which humans evolved, to an earth none of us have ever seen but long for, when humans coexisted with herds of bison, gaggles of geese, coveys of partridge, and herds of deer, and also with flocks of Carolina parakeets so thick they blackened the sky and rivers boiling gray-pink with migrating salmon. They have reentered a primal abundance of story.

After dinner Nels blew on his digeridoo and Hank played guitar. Somebody uncorked another bottle of wine. Lines from a Kunitz poem came to mind, about having tribes constructed out of "true affections" and how our tribes get scattered. That April evening at the end of a sojourn, I was with my people—our wild, wild people—and on my earth—our wild, wild earth.

Huckleberry cobbler was homemade from handpicked, home-frozen berries, with ice cream. It was followed by a sample of Carolyn's home-brandied cranberries, not like a sauce, instead the wine-colored berries whole and heady.

If you have read Nels's work, you can imagine what a happy, soaring person he was, the kind of person who kayaks alone to an unin-habited island, has an experience where a wild deer comes up and touches him, and while reading the story years later in a college auditorium creates the same breathless, wheeling ecstasy. When I spent time with him that month in Alaska, he was in his early sixties, kept his hair cropped short, and wore glasses, so the degree of his wild nature was not apparent at first glance.

One day Nels had taken us out in his boat. He anchored in a calm bay, and we climbed a beach on Kruzof Island. Nels guided us into a muskeg, walking in deep indentations left by brown bears. He showed us the place in his story where the doe had come right up to him.

On the ocean side of Kruzof, Nels nudged his boat into a room-sized cave at sea level, the same as Hank had done, except Nels was less measured. The ocean was rough, the cave room-sized. Nels cut the motor. Behind us the Pacific Ocean swept into the cave in ferocious waves, and before us a merciless wall of rock wept cold tears. The heartbeat of the ocean swept us farther inside, wave upon wave, while Nels, an exuberant Pontus, laughed gleefully. Even with life vests strapped over winter coats, we would not have survived had we capsized.

Finally, as we were about to crash, Nels switched on his engine and backed away.

We turned out to sea. We went on and on. We saw any number of ravens. We saw Steller's jays and goldeneyes. We saw seal, and sea otter, floating with head and feet out of the water, and then a single small arc of humpback whale. We saw the low gray sky, rain again, a sudden darkening. The far islands were shrouded in mist and in all that, all the stories one needed for lifeblood, all the stories to sustain us.

After that, I trusted the world more. I had seen what I had seen. If wild plenitude still existed somewhere, then it was not a myth, and it could exist again anywhere.

PART III   MAGNITUDE

Manatee

WHEN FIVE HUNDRED ARE LEFT

Crystal River, Florida

Everything is jumbled up. The ones with feet are in water although they can't go anywhere. The ones without feet can go where they want. The ones without hands can touch.

It is early morning, November, and I heave out of a guide boat, splashing into the silverback water at the exact place where one of Florida's high-magnitude limestone springs, Three Sisters, rises into a blackwater river, the Crystal. I am wearing a black wetsuit tight as an extra layer of skin and a snorkel mask. I am here to see manatees, which will not hurt me, I am told, and in their curiosity may approach me although I may not move toward them or even reach out to touch one should it approach.

The water is 72 degrees, the November air is chilly, and to keep warm I stay submerged, floating at the edge of Three Sisters Springs. I can't see more than ten or twelve feet in the light-green, silty water. If manatees are below or around me, as the guides promised, I can't see them. I puff through the blowhole of the snorkel, my face turned away from the light, trying to make sense of the furry stumps and uneven topography of the sepulchral river bottom.

Suddenly two monstrous shapes materialize out of the watery

gloom. They are titanic. They are within yards of me before I see them and my heart leaps.

Manatees are vegetarian, I remind myself. They eat water weeds and grasses. Though many times a human's weight, they are entirely benign.

The manatees paddle slowly, buoyant in the water, languidly moving a fin to propel. Both of them head straight toward me. Who would have imagined their curiosity? Who would have dreamed they wanted us?

The manatees glide past, observing me. They note my big glass eyes, my seal-like skin, my shivering body. "You can touch us," they say. In the next moment I break federal law. I reach out and let the body of the second manatee brush against my hand.

The manatee instantly pivots, as if my touch communicated something necessary. It positions itself alongside me, rolls over, and presents its belly. Its fins are like arms, its tail huge and rounded. It keeps looking at me with one comely brown eye. I am pretty sure that the manatee wants its belly scratched. I put my hand there. Its taboo hide is rough, covered with crusty barnacles and slimy with algae, and also soft, mammalian, warm.

That quickly, a manatee is in my arms.

I can't talk. I can't smile. I can't say anything with my eyes. All I have to communicate with is my hand, now softly stroking the manatee's belly, and also one other thing, which I don't have a name for.

After a minute the manatee slowly paddles on. Three others come. They go. Five appear. They all seem to be taking turns checking out the visitors.

Now, below, I can make out manatees resting on the river bottom, in deep greenish silt. What I thought was the pocked bed of the river is a fat knot of sea cows, defenseless and strange. Some seem to lie atop each other in a surreal and merciful sweetness. When one needs air, it lifts lazily through the layered, calcinated, astonished water, sinking again in that slow fall from sparkle to decay.

I pump my feet and glide. I quit using my hands. I glide again and roll. What interests me this morning is a cathedral of spring, how the rays of new sun refract to points in front of me in the clear and unspeakable water, a fluid kingdom. When a manatee approaches, easing next to me, I put my goggled head against its side.

My cheek rests against a series of ridged scars. The propellers of motorboats have cut the animal deeply, more than once.

I blow water from my tube and examine the manatees I can see. Almost all of their backs are scarred.

A mother and calf dispatch from the group below. They dally closer. The calf quits nursing when it sees me. It comes up as if I'm its friend, tumbles next to me, looks me in the eyes, swims, and tumbles. I roll with it, breathing when I reach the surface. The mother nudges her baby away and rubs her body against mine. She rolls. She does this again and again.

The manatees have mistaken me for something I am not.

The mother stops rolling. With her searching eyes she pores over me. She puts her face next to mine, looking. The manatee's eye is a wrinkled spiral.

Immersed in all the unknown, all the mysteries, we gaze at each other. I shouldn't tell you what happened. It's too precious, really, to reveal something like this so randomly. But maybe if I tell you, your life will be touched, as mine was, and some magnetic poles deep within you will align.

I feel the manatee and myself entering another plane. It is wordless and weightless, fluid, beautifully light. A million crystals are sparkling. We are in the world—the human world where an ecotour guide waits on a boat with dry towels and a cup of cocoa—but also another world. There is no word, really, for this place we have come. It is one of the otherworlds, a place beyond reason, beyond the material, beyond the visible. A manatee's spirit is big, and it will merge with a human's spirit, which is likewise big.

Then I hear the manatee mother speak. She is beseeching me. "You must help us," she says. "You must help us."

I hear her distinctly: "You must help us."

She turns, blows at the surface, nudges her baby, and sinks away, back into the descension of the primitive river bottom. Something rises in me that has been rising for a long time, and I break into the sentient air, dizzy, trembling, and blind with love.

Night Life

The night was moonless. When I breathed in, I was sure I'd find my way, and when I breathed out, I knew I'd be lost forever.

I had been running for over an hour, my pupils widening at first to sweep in the last tailfeathers of day, trying to beat night back to the Springer Mountain trailhead, where I had started out that morning with the entire day wide open before me, the sky a blue lens shining through the green swirls of deciduous forest, birds flying like bits of many-colored glass.

All that was gone. After sundown, and after twilight, and after dusk, blackness had fallen. Night had caught me in its net. My body had vanished until I appeared even to myself as a ghost, running through a vanished world. Invisible, I became breath, cold mountain air curling around me. I could sink somewhere in leaves and wait for morning, ten hours away, or I could seek my way through darkness, down the mountain's long backbone.

Springer Mountain is the southern starting point of the Appalachian Trail, which runs from Georgia to Maine. I was not a through-hiker but a day-hiker, seeing what I could see. I had seen trailing arbutus and large-flowered trillium, Indian physic and spear-leaved yellow violet, wild ginger and yellow root. I had seen northern oriole and whippoorwill, the acorn caches of chipmunks, and, under a gargantuan oak, the pellet of an owl. I had seen a deer antler and what looked like the skull of a raccoon.

Now I was caught by night, wondering if I should construct an emergency camp with nothing in my pockets except a purple bandanna and a red-bladed knife, nothing in my pack except the wrappings of an energy bar and a bag of gorp. I figured I had about two miles before me, back to the parking lot.

The plummeting temperatures of early spring made my decision. I needed to find my way.

The path, decades old, was a long groove through the hills. On either long side of it, leaves on the forest floor sang and rattled the winter's dryness. Had the trail been more used, it would have been kicked clean, worn to earth and pebble. But hikers were only beginning to start their treks north along the Appalachian Trail. Two missteps and I'd be lost.

I removed my shoes, to better feel the ground. As I stepped, I allowed each foot to hover, lightly searching the cold and aching earth. Sometimes I went down to my hands and knees, feeling for indentations, for spaces most cleared of leaf litter and twigs. By degrees I kept to the trail, inching forward.

The human eye has the ability to adjust to darkness, its pupil opening wide to gather any available dusting of light. To have night vision requires good eye health as well as practice navigating darkness, in order to exercise the light-sensitive rod cells of the retina. My eyes began to see things it could not see. In the absence of sufficient light, capacity once devoted to vision opened to other sensory receptors. My body became a sponge, highly alert to information, questing to survive.

Years later, I would remember that night as the first time I confronted darkness, to learn what it knew, to conquer it, to pass through it. By midnight I had come out of the woods.

I, like most, was the kind of kid who every night checked under my bed and in my closet for bogeymen. I navigated my childhood home light-switch to light-switch.

When I became an adult, I made an effort to accept the darkness. I quit using flashlights. I shielded streetlights when I lived in the city and disconnected security lights when I lived in the country. I unplugged night-lights, never left door lights burning during an evening away. I did this not only because throwaway batteries and lights burning in empty rooms insult my sense of frugality and my ethos of conservation, but also because I believe fear is at the basis of most night-light.

I have not spared myself. I have crept slowly through the night, arms raised to protect my eyes. I have entered empty, dark rooms, clawing at air, terrified that I will feel a human face or that I will trip over a human limb. I have memorized numbers of stairs, maps to bathrooms, grids of furniture, gauntlets of flower beds and shrubbery.

We humans have told ourselves that day is white, night is black. Day is good, night is wicked. Day is productive, night is frittery. Day is wakeful, night is asleep.

But that is wrong. At night half of the world opens.

Once when I lived in the Andes Mountains of Colombia, I met a man named Miguel who had been born in a small village only reachable by foot. He was ten years old, he told me, when he first saw what he called *luz artificial*, or electric light. By day and by night, as he grew into a man, his light came only from fire, natural and real.

Now I remember a thousand night skies. I remember best the

darkest nights. I remember the Montana prairie when Hale-Bopp streaked through. I remember the Alaska sky, hoping for northern lights. I remember the Perseids in Nova Scotia. I remember the depth of black velvet in the Sacred Valley of Peru. One autumn night at an observatory in Oxford, Mississippi, Mars hovered as close to Earth as it had been in 60,000 years; and another spring evening, as I camped on the Altamaha River, five planets—Jupiter, Saturn, Mars, Venus, and Mercury—bunched beneath a boat moon.

I have judged the quality of a place by its depth of night, so that when I arrive some place new, on the first night I slip outside and look up. The more stars in the sky and the brighter the Milky Way, the happier I am. Sometimes I can clearly see all seven sisters.

I have found out that what we miss, in our love for daylight and things of day, is a nocturnal natural history of fabulous proportion. Spring and fall, birds fill the night sky, mostly unseen. High in the universe yellow-billed cuckoos migrate, their bodies silhouetted against harvest moons. On spring nights, male pinewoods tree frogs hop to the edges of vernal pools, where they plead in synchrony, begging a mistress to come to them. Flatwoods salamanders pick a cold rainy fall night to start their treks downhill to ephemeral breeding ponds. By night, sea turtles migrate toward sand dunes, where they painstakingly dig nests and lay their leathery eggs. Mullet leaping in the Gulf waters throw up spumes of phosphorescence. Owls hunt the understory.

These animals are wired to move in darkness. Light will confuse them, stall them, disorient them.

Might we humans too be stupefied by an absence of darkness? Might we, in eternal artificial light, fail to thrive? Might constant light be a form of torture?

Flying over cities at night, looking down, I used to think, "It might as well be day down there." My friend Sandy West, who lived on an otherwise uninhabited barrier island named Ossabaw off the coast of Georgia, and who died on her birthday in 2021 at the age of 108, talked to me once about her sadness for the modern lives of most young people. "They will never know total darkness," she said. "We have all but completely taken night away from them."

Partly what has confused us is the metaphor. Our desire for meaning keeps us reaching for greater clarity and luminosity. We confound lucidity and transparency with kilowatts. We confuse artificial light with enlightenment. We may be so afraid of darkness that we destroy it, banishing the illumination that darkness brings.

Every year on cold rainy nights in March and April in New England, spotted salamanders leave hibernation to labor across leaf litter and stones, inching downhill out of maple and birch forests toward the ponds and wood pools where they will breed. Thousands upon thousands pause at the edges of roadways, and then begin to cross. Other amphibians migrate too: strange red newts, woods frogs, toads, and spring peepers.

The salamanders are a wet gray, spotted bright yellow. They are as much as eight inches long. Their eyes are bewitching, their bodies fleshy and willing. Some of them are twenty years old.

When I lived in the small city of Brattleboro, Vermont, I got to know these salamanders. During their early-spring breeding season a few hundred of them would make their way across Orchard Street, a dirt road near my home, heading for a cattail pond. On nights they migrated I would go help the animals across Orchard Street; I was one of dozens such volunteers in town. If no one helps the salamanders, a portion of them are crushed beneath vehicles that pass, and the carnage is unbearable.

It was late March. It was raining and finally above 40 degrees, and spotted salamanders were pouring downhill to the vernal pool to reproduce. One crawled out of the upland woods into my beam of light. I bent and picked it up. It lay coolly black, beautiful in its yellow polka dots, in its endearing bulging eyes a hapless, vulnerable look.

I put it down. What the spotted salamander wanted was to reach the sedge-rimmed pond and lay its eggs. As long as there were no cars, it could get itself across. I made a mark on my paper.

Up and down the migration zone I scouted, stepping carefully, noting salamanders as they crossed the road, counting.

When headlights approached, I stopped the driver with a wave and a smile, explained myself, and guided the car through the zone. The pond was filling with frogs and sounding crazy.

The night went on. There was less traffic, which was good, since more and more salamanders were making their way through the woods downhill. I switched off my light and stood in darkness. The dark wood collected around me. Though the frog chorus from the pond was deafening, I could hear tiny rustlings of leaves, the litter crackling as salamanders crawled through it on their bellies. Around me hundreds of them made slow, patient, fierce progress toward breeding grounds. They were small, thin-skinned, delicate. The rain, as it fell, tinkled like small and constant kisses. A barred owl called.

Standing, listening to them, I felt my body become the earth's body. Spotted salamanders crossed my body to get to the cool mysteries of the pond. All that I could offer them I offered—warmth, sustenance, home. I was the hillock. I was the maples and birch growing there. I was the placid bowl of pond, vibrating.

I dared not move then, for fear of treading on some life. I stood for a long time beneath night clouds, looking into the inscrutable distance. Life flowed and flowed past me, like a river, until it finally lifted me in its insurgence and delivered me to the edge of a luscious and telling obscurity.

A Terrible and Beautiful Scar

ON SLAVERY

Near Tallahassee, Florida

If instead of a human nature, I could choose a wild one, it would be a southern blackwater stream, and of all of them I would choose the Slave Canal.

In the karst limestone of north Florida, rivers rise through high-magnitude springs, then vanish. Their swirling waters dissolve soft rock to create a mostly hidden realm of caverns, pools, and passages. Near the town of Miccosukee, the Wacissa River emerges through a series of clearwater springs only to disappear into braided channels through the swamps of that territory. It is only passable, then, for a short distance. Not far east the Aucilla River rises, and it dips below ground at crystal-clear springs before becoming tidal in its rush to join the Gulf of Mexico.

Across this landscape a gash called the Slave Canal is laid, three miles long, hauntingly beautiful, a wounded soul journeying through gloomy forests of cypress and tupelo.

In the early 1800s John Gamble owned a cotton plantation in Jefferson County, Florida. The river closest to his plantation, the Wacissa, was impassible in places, so Gamble was forced to haul his cotton overland to the lower Aucilla in order to ship it downriver to the Gulf. Three miles of unnavigable swamp separated the Wacissa River from the Aucilla. If Gamble could dig a three-mile canal through that swamp, connecting the two rivers, he could save himself eighteen miles of road travel, I'm told.

On December 19, 1850, Florida's General Assembly approved Gamble's plan to make the Wacissa and Aucilla Rivers sufficiently navigable for flat-bottomed boats. Enslaved people would do the work.

We will never know how many workers lived and how many died in the construction of the Slave Canal. Someone told me the number was in the hundreds. But that figure would not have been economically feasible, maybe not historically possible.

Even if no person died, it was holocaust in the land between the rivers.

My first trip through the Slave Canal, many years ago, I could feel the dark sorrow of the laborers. I imagined them plunging shovels and picks into black mud, picking up rocks as if they lifted their own heavy hearts, bearing their heaviness to shore again and again and again. I saw brown and black men bending in the dimness of the bottomland hardwoods, at the edge of brown and black water, boulders of limestone in their arms. I could see them wrestling among the thick, deep roots of the cypress, watching for water moccasins. Mossy limestone lined the canal banks like the torsos of men. I could hear old spirituals and small moans and whispers and sometimes even laughter.

When I paddled the tannic waters, the lonely dead softly sang, *Go down, Moses*, while white-eyed vireos flitted through the trees. The dead muttered among each other as I passed, so that the sadness of the Slave Canal settled on me and clung to me like tobacco smoke. In my boat I rowed the endless tragedy of racism, a heartbreak I've felt every day since I became cognitive of skin color. I was a toddler then, watching the way my people spoke differently to other people, watching the way faces changed in the presence of the other, watching where some people could go and where some could not go.

The waters of the South are dark.

My dark soul runs with the waters.

We say of the Slave Canal that we go in and come out, as if doors hinge at either end, curtained with vines—Virginia creeper and wild grape—that drape through river birch. We think of the watercourse as a passage, a rite of passage, built by an aristocracy who wanted their right of passage. It is a river of sticks, River Styx, where Michael rowed his boat ashore, hallelujah. The canal is a scar through a cabbage-palm hammock, a wound gouged deep that healed but whose story will never be forgotten. It is a laceration drawn in the 1850s on the map of the world, an unforgivable mark of history.

No, let me change that. It is a *forgivable* mark of history. This particular horror is in the past, which dives backward in time, back and back and back in layers of history—far, far, far. We all come from oppressors and also from victims. We all are the products of aggression as well as kindness. We have all smote and been smitten.

Pray tell me, the "better angels of our nature" that President Lincoln wrote about and the arc of human nature that Dr. King spoke of surely indicate a beatific movement toward luminosity. Surely that would include reconciliation.

I often contemplate this idea of forgiveness. Something happened to someone I deeply love that I have found unforgivable. It *was* unforgiveable. What happened was a form of slavery. I can't say much about it, except that, as slavery ruined the "pursuit of life, liberty, and happiness" for enslaved people, what happened to this person I love ruined her life, permanently.

My favorite priest told me that forgiveness was a gift to myself, not to the perpetrator. She told me that forgiveness was not *forgetting*; it was refusing to be diminished by ill feelings toward another. I still don't fully understand. My hatred for the perpetrator was so impenetrable that even when he unexpectedly died I could not find a speck of forgiveness in my heart for him. I wanted him to be resurrected so that he could die again. Disposing of the ashes of this person fell to me, and I could not think of a place that deserved such contamination. I weep admitting this.

I say it to admit that forgiveness is difficult, sometimes impossible. I hear that. In terms of the bigger picture, I think what I am trying to say is that we have to take great care, when naming an oppression and an oppressor, to identify the correct one. We have to be careful to name a current oppressor for a current oppression, and to work to stop it, in ourselves (if we are the oppressor) and in others—to force it to be stopped.

Perhaps no word exists in our terribly inadequate English language to name this abstract, emotional thing that is not forgiveness, is not forgetting, and also is both. I didn't feel it on the Slave Canal. What I felt there was bottomless sadness.

The Slave Canal does not tend its sorrow. None of its animals tend it. The barred owl watches me paddle by, dipping my blades back and forth. The owl's grandmother many times removed watched African American men laboring waist-deep in water and mud.

Such beauty. Cottonmouths embrace the branches of fallen trees. Necklaces of cooters sun on angled logs. Flags of whitetail deer fly off through the woods. Yellow-billed cuckoos clatter. Prothonotary warblers, also called swamp canaries, dash about like daredevils through the humid heavens. The shallows thicken with water hyacinth, blooming, and with royal fern, sweetening the eye of the alligator. Occasional clumps of cardinal flowers beckon hummingbirds.

I have never lived a moment not yearning to undo the damage.

In August spiders build elaborate webs across waterways, where they crouch with infinite patience and wait for fliers-by. They crawl along the branches of beech, they leap from thin air, they hang by threads. There are crab spiders and banana spiders and zipper spiders and so many others I do not have the names for. In morning sunlight the webs of golden orb weavers look like tapestries of gold filigree. The river is bright with all the gold.

If spiders own the trees of the Slave Canal, butterflies own the skies. Tiger swallowtails as big as my hand flit through the upper reaches, landing on swamp lilies. Palamedes swallowtails, Carolina satyrs, pipevine swallowtails, Gulf fritillaries, and checkered skippers fly into the beautiful golden nets of the spiders and lie trapped until they die, leaving their life stories written in the obituaries of loose wings.

The banks are close, cypress and tupelo and red maple growing to water's edge, leaning over each other to touch the other side, as if the trees know a canopy is supposed to be unbroken. They want to heal. So do I. I dawdle, drifting, watching bluets rumba along the tips of duck-potato and counting apple snail eggs clinging to pickerel weed.

I used to think that there was so much love on earth that the human form, composed as it is of corpuscles and atoms, could not contain it and so it overflowed onto the land, which became a vessel for affection. Now I don't think so. I think the love is in the land, in the black water, and we go to wildness to glean love, to be nourished.

When human savagery and brutality erode me, I long to be on the Slave Canal. I trust the dead. I trust them to open the door and let me in, open the door and let me out.

Cabbage palms rise in the floodplain. Poison ivy vines, thick and hairy, snake up the trees. The blades of wild rice sharpen against my skin as I glide through a kayak-wide channel. Red-shouldered hawks cry in the distance.

I let my eyes fill with the innocence of green. I hear our right to speak in tupelo leaves and I hear dignity in water dripping off my paddle. I see freedom in the sandy-bottomed water. I hear justice in dragonflies as they clack and buzz. I feel on my bare arms the forgiveness of the yellow sun.

What better nature could one have?

Forms of Rarity

LUST FOR LIFE IN FOUR SEASONS
Southeastern United States

Lust drives much of what we do. It lures us toward sexy and iconic natural spectacles such as breaching whales, overwintering monarchs, and congregating manatees. Just as satisfying are small spectacles underfoot that are easy to miss, especially if no guide appears to whisk back a curtain, revealing a world unseen, unknown, and unimagined.

I. SPRING
Breeding Pond, south of Tallahassee, Florida

Somehow I have wrangled an invitation to a vernal pool from a biologist studying pinewoods tree frogs. I follow him through a darkening wood toward a deafening sound of frogs, until a small pond pulsates before us. To the west the last moments of sunset limn the horizon in soft peach, visible through black silhouettes of longleaf pines. Already the pond reflects the unspendable silver of a half-moon.

Without Kit, my guide, I wouldn't have seen this. I wouldn't have been wandering the woods south of Tallahassee, Florida, at night, trying to find pools of rapture.

The pond has no name. They call it Twin Pond South.

Biology students Dave, Amy, and Dina are helping. Mosquitoes are bad so the first thing the researchers do is put their things down and slather with repellant. Kit, who's wearing rubber boots, says we're going around the entire pond putting up a knee-high plastic fence that must be taken down at the end of each night.

I can barely hear him, the frogs are so loud, croaking singing rattling drumming, playing combs playing banjos playing horns thumping bass, an eerie endangered ensemble. Up close one frog pumps his throat sac in and out, earnestly, harmoniously, intently.

Any pond on any spring night can be a world-class orchestra of extraordinarily bawdy proportion. This has been a warm day after heavy rain. Frog noise doesn't get any rowdier than this. I hear sounds like a person rubbing a balloon.

Kit is yelling at me and I am trying to understand him. "Leopard frog," he says. That must be the percussionist of balloons.

A tiny frog lands on my arm. Kit leans over. "Cricket frog. That's the one you hear *tsk*ing." He mimics it, *tsk, tsk, tsk,* and for a minute he's just another horn. Across the pond the lights of the students jump along, reflected in the water. I watch a frog sitting in grass at the edge, belly flat. When it calls, its vocal sac inflates.

Kit has a grant from the National Science Foundation to investigate the frequencies of pinewoods tree frog calls. Almost all females care about some frequency, he says. Larger females may prefer lower frequencies. To find out he plays different frequencies through a sound system and notes which frogs arrive.

Trouble is, even with all this amphibious bounty, pinewoods tree frogs are rare. Our job is to walk around and around the pond looking for them.

"Males come in earlier than females," Kit says. "Most females are in by 9:30 or a quarter to 10, although some are as late as 11. If it's cloudy they come down earlier."

We start walking. Now the fence is up. We're seeing plenty of frogs, but not the endangered ones. We're not getting anybody.

"I feel them coming," Kit says. "They sit around and finally somebody starts calling, then others join in." We complete one revolution and begin the second. By 9:15 we have worn a path around the pond. I wonder how many miles collectively have been spent looping it. At 9:30 there are still no frogs of the species we're seeking.

Finally, from a tree Kit hears a deep, booming *teet*. From a nearby tree another resounds. "That's it," yells Kit. The pinewoods tree frogs call and respond, on and on. "When they get out of sync, they stop and get back in," says Kit. He gets out his computer.

What happens is that the males call. The females come and bump into the males. Then the male climbs onto the female's back in an act called amplexus, a grasp that stimulates the female to release her eggs. She carries him around for four to six hours. During this time she will lay many clusters of eggs among vegetation in the ephemeral wetlands, as many as 2,000 eggs in a cluster. The male fertilizes the eggs externally. Some females mate up to four times, says Kit. Matings are separated by three weeks on average.

Males produce fifty-five calls per minute on average for 2.5 hours, although they've been known to go for 4.5 hours. That could add up to 8,000 calls per night. Muscles in the males' dark-green throats force air in and out. "It's energetically expensive," Kit says.

Carrying your mate on your back for four to six hours is energetically expensive too, I'm thinking, as is laying thousands of eggs.

Plus this is a dangerous place. Bellowing, single-minded males become really obvious to predators. A water snake cruises every night munching on them, the researchers tell me. The chance that any one male is killed is 5 percent.

Now the computer begins to generate synthetic calls, adding its own bellowing to the crazy cacophony. Kit is trying to attract female pinewoods tree frogs. When they show up, he will grab them (sometimes they squeal, he says) and measure them. "Females will be absolutely convinced that there's a male here," Kit says. "They will hump the speakers." Sometimes the frogs jump from the trees to get into the bedlam. "I almost got hit on the head once," says Kit.

I'm going crazy with all the noise. I want to clamp my hands over my ears. How do the frogs find each other with all the heeing and hawing, tisking and tasking, hooting and hollering?

A human is calling now, one of the students. "I've got one at fifty-four. Want me to bring him over?"

"We'll come to you," Kit shouts.

We move around the pond, lights on foreheads. Outside the perimeters of our beams the ecosphere is pitch-black.

The frog is jade-green with russet spots and stripes down his short legs. It's a tiny and strange amphibious form with a profound joie de vivre. I don't touch it. The male is on a mission, to repopulate its domain and range with more frogs to keep the centuries of evolution evolving.

The drive to mate is a powerful drug. High on it, frogs will bumble their way through the flatwoods to ephemeral ponds like this one, they will leap from high in trees, and they will hump speakers. The collective racket of their desire is deafening. Somehow among all the frogs of all the species they will find a mate, then the female will lay her eggs in the water, sometimes until it becomes milky. This is how life goes on. Life requires this raucous, rollicking desire. When a person like me gets to see it and hear it, the feeling is also a powerful drug. Every living thing is high in the flatwoods on a spring night as the frogs do what they have done forever.

The night wears on, and Kit and the students do what they do. The computer calls, the females come, Kit measures them, lets them go. I watch it all as if in a dream, thinking how rare the natural spectacle is and how it brings a human like me as close as I've ever been to ecstasy.

II. SUMMER

Wakulla Beach, Florida

This morning I have come to Wakulla Beach, about eight miles from the house, to think about the world, where wars are raging. The long sand road to the beach was deeply rutted by recent rains from Tropical Storm Isadora. Two trees were down across the road, and someone had chainsawed a car path through them. Mostly fishers launching johnboats from the dirt ramp come here.

I sit secluded in a grove of gnarled and leaning cedars, looking out across a mudflat and a saltmarsh. It's newly fall and a light, variable wind keeps the cabbage palms rattling their fronds. Above, a lone cicada saws her wings together, playing on her fiddle the only tune she needs. A monarch butterfly, which will soon join hundreds of others for their pilgrimage across the Gulf, drifts along the tree line. A clapper rail hops out of the branches twenty feet from me and eases toward the mudflat, where she begins to feed on fiddler crabs.

I am not a statesman, politician, economist, or oil magnate. I am neither fighter nor chieftain. I am only a subsistence farmer and a poet. What I want from this life are simple things. What I want is to sit motionless, as long as my body allows, and watch the rail's eye, which is the burning red of the sun. What I want is to know my place the way the bird knows hers. The rail does not think of war. She can be strong without owning the rest of the world. She shows us that power is being in charge of our own lives.

My life has been a long movement toward strengthening family, community, and land, and also intuition and spirit. This is all the security I need.

I can tell you about peace—this peace, today. A great egret wings over the marsh, black legs straight out behind her. Another bright-orange monarch flutters by. A paddle hits the side of a boat. The wind murmurs.

III. FALL

Okefenokee Swamp, Georgia

The world is unaccountably beautiful. In fall in Okefenokee Swamp, grasses flower in the savannas. I stand in my boat to gain a better vantage, poling through shallow creek runs like an old-timer, looking out across the magnificent flats, where asters and yellow-eyed grass and meadow beauty bloom. I sing softly.

Tiny cricket frogs hop from lily pad to spatterdock, away from my approach. A pair of blue-winged teal, surprised, take to wing, circling the wet prairie. A yellow-billed cuckoo ducks into a cypress dome. In Big Water, a doe wades out to drink.

I am traversing the watery wilderness, a three-day trip of thirty-one miles, this time with my husband, Raven. The 9/11 attack has just happened: the Twin Towers have fallen, and 3,000 people have been killed. Somewhere, bombs are dropping from the sky, each one inscribing a circle of destruction. Somewhere someone is planning to kill someone. Someone has been killed. Someone is grieving a death. Somewhere someone is hiding for her life.

But not here.

Here, I am far from terror. The nearest city, with its susceptibility, is a hundred miles away. I have seen only Raven, no one else, since we left Kingfisher Landing two days ago. In this swamp there are more sandhill cranes and more black bears than people, and I am afraid of neither the cranes nor the bears. I am not worried about a death I have not yet imagined.

detritus—plastic buckets, Styrofoam crab floats, rubber gloves, plastic drink bottles, aluminum cans. Part of a dock has washed ashore, treated lumber with nails exposed. The things we manufacture and use away from this wild refuge wind up here anyway, in the one place they should not be.

I stand and look out at the Gulf, my head full of terrible thoughts, thinking about an oil spill, the one that has not yet happened but will, and what might happen to this wild island and its inhabitants. The thought of it is too horrible to dwell on. The worst of human civilization insinuates itself in the best of what we have left.

I come upon a path and I turn inland. A faded sign says "Eagle Nesting Area," but the eagles are not nesting in winter. The maritime forest is alive with magnolia and live oak, also with tall cabbage palms. This is not a thin forest, not a sparse forest, not a new forest, not a forest to be taken lightly. This is an old forest on an old barrier island where red wolves have been calling.

As I walk I am remembering the young woman I used to be, who came alive in windswept geographies. I am coming back from the dead. I am far, far away from everything I don't like about the world, deep in the arms of what I love most.

These feelings are especially strong because the Gulf of Mexico was the territory in which I came of age, after I arrived as an undergraduate in Tallahassee. Here I first saw plovers nesting on beach sand. I saw a freshwater spring bubbling from the salty depths of the Gulf. I saw dolphins leap. I made my first bird list and retrieved my first scallop, caught my first shark. I tasted smoked mullet and found ancient pottery shards and identified an oystercatcher. Here, as I said, I came alive.

Along my path now I chance upon some enormous bones, pelvis and tibia, which must have belonged to a sambar deer, an introduced species. In a chalky owl pellet (its location revealed by white stains on leaves) I poke around and find the tiny skull of a rodent. I am happy in the old coastal woods, happy in the middle of my life, happy in the wildness, happy in this moment.

As I walk I am making resolutions.

In the New Year, outside will be inside. I will spend more nights out of doors. I will canoe more rivers. I will climb more trees. I will celebrate the high holidays of the sun, earth's calendar, with ceremonies that involve fire and not money. I will pay more attention to birds. I will attempt to be the kind of person animals want to chat with. I will learn the names of more plants and I will listen to them.

I will use less, stay home more. I will think about the consequences of my decisions on even the smallest menhaden in the sea. I will listen better to what the earth is saying, and I will write it down. When it needs help, I will help it.

Spiderwomen

What place is more bereft of life than a cemetery, filled with the dead as it is, paved with stones, and planted in turf? So I thought until one early-summer evening in the churchyard of St. Peter's Episcopal in Mississippi.

I stood looking into a bowl of St. Peter's Cemetery. Below, a small grove of ancient cedars grew in a circle, druidic and consolatory. In the salmon-colored sky thunderheads were building, lightning not far off. In the next two hours I would see the sun go down, dusk fall, and the graveyard fill with fireflies. Their blinking lights would overflow the bowl, like sparkling cream spilling over.

That was not the amazement.

No, I would see among the slabs and headstones planets of spiders living spectacularly complicated and richly inconceivable lives. Two arachnologists would open my eyes to a spider underworld.

Pat Miller and Gail Stratton both taught and conducted research, Pat at Northwest Mississippi Community College and Gail at the University of Mississippi.

Pat, a slight woman with straight dark hair bobbed short, is a native Oxonian. She went away, like kids do. What motivated her return were spider range-maps that showed no dots in Mississippi although she knew Mississippi was full of spiders.

Pat decided she was going to list all the spider species of Mississippi.

Gail, slight and dark-haired but with curls, is as crazy about spiders as Pat. She wants to learn everything she can about them. So far Gail has described two new species, both of them wolf spiders, one of them found eight miles from this cemetery, collected one dark night in forest leaf litter and named *uetzi* after Dr. George W. Uetz, her arachnologist mentor and friend.

Somewhere along the way the two arachnologists fell in love, a beautiful web of spider love, merging their dreams and goals and plans. Every time I've seen them they've been together, conjoined and seemingly inseparable, and of a piece, partners vocationally and avocationally. They are so close they seem to move and speak and think as one, and I think of them that way, as one.

I made the same mistake with spiders. Every spider is just a spider. But no. There are minuscule variations in spider species, differentiated by such things as tibial bristles on the first pair of legs or length of the tiny carapace. Getting to know spiders requires careful, yearslong investigation, an enterprise I don't have the passion for, although I do lust to know some of what the spiderwomen know.

Gail and Pat are very different people, of course. One is quiet until she's talking spiders, then she bubbles. One enjoys small talk. One smiles more than the other, who also smiles a lot. One may be a few years older, I'm not sure. One has straighter hair than the other, as I've said. One has more delicate skin. One has darker eyes.

As a reporter, I should be writing down who said what. Mostly I don't. If one says something, I think of both of them as saying it.

One finishes the other's sentences. The other adds a phrase, completes a thought, makes a little hum of agreement.

I expect that we're going to wait for darkness to shine flashlights around the grave markers looking for the red pinpricks of reflection that indicate spiders. I've done this before—gone out at night with some scout or other to look for orb-weavers in dewy grasses. But we are not doing this.

Few spider families have eyeshine, the spiderwomen tell me (meaning that light does not bounce off their tapetums, the layers of reflective cells at the backs of their eyes). So the spiderwomen set off wandering through the graves, scanning for tiny movements and other patterns I don't see. They say you need to develop a search image, which you hold in your mind, and that image helps you to see and recognize a particular shape.

I don't really understand this. Are they looking for moving dots?

"Sort of," they say. "We just know what we're looking for."

I know what I'm looking for too. I'm looking not to have a spider crawl up my britches.

I tail the arachnologists, asking too many questions. Patient and painstaking, they stop and answer each question as best they can. How clunkily a neophyte enters a world whose immensity she has not yet realized.

"Are we looking for really tiny dots?"

"Sure," says one of the women.

"How small?" I ask.

"There are spiders so bitty they can get stuck in a droplet of water and drown," says one of them, I think Gail.

My mind makes that leap into empathy—something so tiny it drowns in a drop of rain. The world seems so cruel, even when we're talking about spiders.

"And you look for bigger dots," says Pat.

"How big?"

The women stop and turn, both of them. They are so patient that I am embarrassed.

"The largest spiders here in Mississippi are fishing spiders," one of them tells me. She turns to her wife: "Remember the one we found on the boardwalk at Sardis Lake?"

"It was sitting there looking at us with huge eyes," the other says.

"It was intimidating," says the first one. My mind leaps again and goes sailing off to a boardwalk at Sardis Lake where a spider the size of a raccoon is looking up at me.

"How big is a fishing spider?" I ask.

"They can stretch out to three inches," she says.

I decide to stay quiet a while so the women can do what they're supposed to be doing here. Darkness is coming fast and the thunderstorm is closer.

One of the women spots a spider, then another, then another. The other spots a spider, then another, then another. They want to show them to me and I want to see them. They murmur multisyllabic Latinate names, followed by an observation or two, and most of this skyrockets off into the fireflies above my head.

The spiderwomen have eyes for spiders. They could find them all night, and the night would become a list. I realize the endlessness of this exercise, with 550 species of spiders identified so far in Mississippi and about 49,500 named in the world, according to the World Spider Catalog. (Closer to 170,000 are believed to exist.)

Tonight they're looking for a certain rare *Schizocosa*, a type of wolf spider, which they've found three times before.

Wolf spiders are common in many habitats throughout North America. They have a stout body shape, their eight legs are fairly even lengthwise, and their eight eyes, which you may need a hand lens to see, are arranged so they can see above and behind them-selves. Most wolf spiders don't build webs; instead they burrow in the ground or hide in the leaf litter.

They've found the *Schizocosa* in this cemetery. They found it one Mother's Day while cleaning Pat's mother's grave.

"My mom is buried here," Pat says. "Over there."

"We'd just like to find it again," says Gail. They keep looking, murmuring quietly, *Here's a this* and *Here's a that* and *Have you ever seen a such-and-such? Here's a pair mating. This is what a burrow looks like. That's a pretty big burrow.* Their flashlights make little streaks of light, while in the sky the lightning has started to make big streaks.

I ask them what all reporters ask their subjects: Why? Why do you wander around weird places at night with flashlights, looking for freaky things?

Gail looks at Pat, who is quiet. Gail aims her light at the ground. "What motivates me is how fascinating—how amazingly com-plex—spiders are," she says. "It's humbling." She talks about spiders being able to learn. A female was exposed to male spiders with dark legs, and after three weeks she preferred dark-legged spiders, al-though unexposed females preferred light-legged ones. *DA-yum,* I think.

She goes on. "Did you know spiders have a song and a dance?"

I didn't know this, and I don't think I want to know. The more I find out, the harder it's going to be to smash one of these babies when it builds its web above my bathtub. I still haven't recovered from the raindrop image.

"Yes, it's true," Pat says. "We're recording them."

The spiderwomen are recording spiders singing and dancing. This seems especially zany here: we are in Mississippi, home of the blues, where music is fundamental, where at the Thacker Moun-tain Radio Show I've heard jug bands, gospel choirs, blues singers, harmonica desperados, guitar geniuses. My limited imagination does not allow me to fathom how a tiny spider song, out of range of the naked ear, or a tiny spider dance, invisible to the eye, could be captured on tape.

"Their sound is both airborne and seismic," Gail (I think) says. "Their world is a vibration world. Spiders 'hear' through something called a slit-sense organ in their legs, which is a kind of strain gauge."

"How do you record it?" I ask.

"Three ways," Gail says. One is a phonograph needle through an amplifier, which picks up sounds on a substrate, whatever that means; another is a sound box with a microphone in the bottom; or, if your lab has lots of money, there's something called a laser vibrometer.

I'm on a science expedition to a local landmark, a cemetery in Oxford, Mississippi, as night falls and a storm bears down, listening to spiderwomen talk about spiders singing and dancing.

"Why do they do it?" I ask.

"Courtship ritual," says Gail. "A male dances in front of a female. If the female's feeling receptive, she'll respond by slow-walking, stopping, turning, then settling down."

"The male approaches carefully so as not to be eaten," Pat says.

The males dance in various ways to produce different sounds. They make intricate moves—flutters, double waves, leg taps. Sometimes they smash their abdomens against the ground, which records as a boom.

"There's an order to the dancing," Pat says.

Gail continues. "They also have a scraper and file, like a little washboard."

The arachnologist couple was filming spiders when they kept hearing somebody outside the building trying to start a motorcycle. They shot exasperated looks at each other. They had posted signs that they would be doing high-level recordings, asking for people to be quiet. "How rude can that person be?" one said. "Don't they know we're filming?" and "I wish they would get the thing cranked and go on."

The motorcycle went on and on.

Finally one of them realized the sound was coming from the spider, an undescribed species they'd found in Homochitto National Forest in southern Mississippi. They joked about naming it *Schizocosa harleydavidsoni*.

Now it's fully dark and we're hearing thunder. Lightning is closer, flashing in the sky behind all the lightning bugs.

"Maybe we should go," one says.

"Well, you've ruined it for me," I say.

Both women laugh. "How so?" says Pat.

"I'm not sure I needed to know about all this. I'll never be able to look at a cemetery the same way."

They laugh again. "It's not just cemeteries," they say. "It's everywhere."

"Exactly," I say.

Inside this world, invisible except in moments, spiders by the millions live their lives in nooks, crannies, niches, and cracks, their world like a spirit-world but not. Their world is a real world, a universe inside the multiverse. Two women in a Mississippi cemetery shine tiny lights into that big dark world. One evening they showed it to me.

On the way out the raindrops were like cold fat stars falling on my face.

POSTSCRIPT: After forty years studying spiders, Pat and Gail have both retired. Gail has become a Unitarian Universalist minister. When Pat retired, she took a mycology class and has now built a mycological collection at the University of Mississippi with over 2,000 specimens that are measured, photographed, spore printed, and cataloged on a searchable database. She and Gail go out a few times a week looking for mushrooms and are again amazed at the diversity. Pat wrote me recently: "Our hair is gray now and I turn 70 this year. It's not so easy to stoop over to look at spiders, and our eyesight is not what it used to be. I have as much passion for mushrooms as I had for spiders, plus they are easier to study. We are still inseparable, and being in the woods has kept us sane through this pandemic." Gail added, "We were saying just today how great it is that we love to do the same things."

I Have Seen the Warrior

HOW NOT TO DIE

Okefenokee Swamp, Georgia

When crossing the largest swamp east of the Mississippi, a three-day journey through 700 square miles of wilderness full of alligators, the goal is clear: don't lose anyone.

We started out with nine—three women and a man, plus five boys about fourteen who acted like boys do, which is goofy, anything for a laugh, teeter-tottering between X-men and grown men, our sons and their friends.

The boy Zachary was riding in my canoe. He was dark-haired and dark-eyed, tall for his age. He was big-hearted but more aloof. Sometimes he didn't like the antics of the other boys. He got angry more quickly. Even then I could see in him the honest, dependable, loyal man he would become.

Zack lived at the very edge of the swamp and had been in many times with his father, Jackie, but had never paddled across it. I was good friends with his father. Jackie was a swashbuckling alligator trapper, bear hunter, and coyote tracker. He was made more renegade by a stint in Vietnam, where one ill-fated Easter Sunday going to church he stepped on a land mine that left enough shrapnel in his body to set off metal detectors.

Jackie would have drunk himself to death, he told me, except the swamp saved him. He would go in for two weeks at a time, camping on some island or other, watching sandhill cranes and wood ducks, following bear trails, drinking. The swamp was his salvation, his refuge in time of trouble.

So I was surprised at what happened. Years later, I still think of it.

I believe death is the ultimate wilderness, waiting to be explored. I don't want to go there yet. As a lover of nature, I want to get as close as possible to essential wildness without actually dying, so I'm careful to keep on life's side of life, skirting the trackless territory of death and mitigating the forces that press us toward it.

I had assembled a first-aid kit, imagining possible attacks on the human body—cuts, scrapes, punctures, splinters, sunburn, wasp stings, chigger digs, flea pricks, snake bites, wildcat rips, giardia, viruses, bacteria, yeasts, constipations, diarrheas, muscle aches, appendicitis, sinusitis, pancreatitis, tonsillitis, heart attacks, liver failures.

We might need Band-Aids and antibiotic cream, rubbing alcohol and aspirin. I put in a tourniquet, although I'd never used one and might not even know how. A stethoscope, being analytical and not curative, wasn't useful. A defibrillator was too heavy. Antivenin was not affordable. I decided against packing the *US Army Survival Handbook*.

Zack and I were in the water first. He was in the bow.

"Can I have a drink?" Zack said.

A lot was happening, and I didn't look back. The others were trying to get their boats launched. "You can have whatever is in this boat," I said.

In seconds Zack gagged and started spitting over the side.

I looked back. "What's wrong?"

"That drink is ruined," he said.

"Oh my god. Zack. That's not a drink."

He seemed confused. "What is it?"

"Lamp oil. Did you actually drink it?"

"I took a big swig," he said.

"Did you swallow any?"

"I swallowed some."

"That's poison. Didn't you see I wrote 'Poison' on the bottle? I drew a skull-and-crossbones on it."

He ducked his head and spat more. "All I knew was I was thirsty."

"We've got to get back out."

The canoe scraped and I stepped into shallow, amber-colored water. The large container of oil was in my truck. Earlier I had grabbed a small Gatorade bottle and portioned a reasonable amount into it. Sure enough, it was poisonous. On the label was a number to call.

I had no experience with poisons. Nothing in the first-aid kit was going to help. I seemed to remember charcoal being an antidote, and milk, but each poison might be unique and have a singular cure. Why didn't the oil container announce the curative in bold letters?

I borrowed a cell phone and called Zack's dad. He thought Zack would be all right. Probably he didn't drink enough to hurt him, he thought. If you called the hotline, you'd be waiting a long time and you knew what they'd say.

"I'm worried about you going," I said to Zack. "How do you feel?"

"I feel fine."

He seemed okay. "If something happens, we can't get back easily," I said. "Maybe you should stay."

"I spit most of it out," he said. "I don't feel anything."

Everybody was waiting now. "Well…let's give it a shot."

I spoke to my own son then, making sure he was ready and had everything he needed. Silas was tall and slim, black-headed, a fun-loving kid who all his life so far has valued friends as his most cherished possessions. He wasn't jealous of Zack riding in my canoe. He had his own kayak and often paddled with me. He'd rather be with his buddies, in their kayaks, than with his mom.

Zack didn't have a mom in his life or experience in a kayak. I enjoyed his company and I wanted to spend time with him. We were in the right boat.

We left Kingfisher Landing headed toward Carter Prairie.

Okefenokee Swamp is a large bog, a depression, a peat-maker, a mosaic of peat batteries—a saucer, not a cup. There is very little solid high ground, only floating sod. The sphagnum that defines the bog is so thick it appears weight-bearing, but when you step onto it, you sink, floundering in the detritus of centuries of vegetation. The Okefenokee is called "land of the trembling earth" for a reason. Therefore, a prairie in the swamp is not what you would think. It's a flat, floral wetland.

The day was feverishly gorgeous. Carter Prairie was popping with water lily, spatterdock sticking up its yellow bonnets, swamp iris in amethyst and plum, golden club like birthday candles. In shallower places carnivorous plants were going crazy: yellow bladderwort, purple bladderwort, and both the hooded pitcher plant and the golden trumpet, blooming their flappy flowers. Hatpins were everywhere, like tiny marshmallows on sticks or white balloons on strings or little flags on delicate poles. All the metaphors I can think of for hatpins are happy.

The water was mica-colored, lava-like, obsidian glass, reflecting white puffy clouds in an indigo sky. The trail flecked with foam. Prothonotary warblers called, *whk whk whk whk whk*. Dragonflies, some golden reddish and some powder blue, zipped along. Startled kingfishers careered raucously away.

The first night we'd stay at the shelter at Maul Hammock, twelve miles in. The second night we'd stay at Big Water. By the time we emerged at Billy's Lake on the western edge, we'd have traveled thirty-one miles.

When most of us think of wilderness, we think of Alaska or the West. But there's wild wilderness even in places like south Georgia that have been inhabited for long centuries and degraded in the process. For me, not enough wilderness is left, and every day I resent this. It's a crying shame that humans have trampled so much of this world with such abandon.

Every guidebook we'd read said the twelve miles to Maul Hammock were hard miles. Very soon we saw why. In a flat there's no current; only sheer muscular force will propel a boat. Sometimes we had to get out and drag our boats through the dinner-plate leaves of water lily and spatterdock, tangles of flowers: pulse-revving and hypnotic but exhausting.

Zack maintained his share of paddling. I kept asking him how he was doing and he kept saying he was good. He seemed good. Maybe he got the poison out. Still, I was angry at him for not paying attention and angry at myself for not ripping off the Gatorade label and for not hiding the bottle where a kid couldn't find it. I encouraged him occasionally to eat crackers. I thought they would settle his stomach and absorb lamp oil residue.

We passed mile markers #1 and #2 and #3. Once we stopped to knock our boats into a knot and pass around energy bars. At #4, where a dark-water pool was ringed with trees, we tied up to the wooden marker and ate lunch.

"Can we swim?" one of the boys asked.

"Not a good idea," said one of the adults. The gators had pressed wide highways through the peat battery around us.

The boys found a log up-slanted out of the water. One by one they crawled onto it, careful to stay balanced.

I was responsible for one other kid, Max, one of Silas's school friends. He loved to test limits in the most endearing ways. He spent a lot of time at my house and I knew what to expect from him.

Of course it was Max who slipped and fell in. He surfaced, brown eyes large above the water, thrashing.

"Ayyyyy," he shrieked. "Get me out of here."

Here I admit to laughing. Max churned back to the tree and tried to pull himself up. He slipped again and disappeared underwater. He surfaced.

"I'm gonna die!" he said. Even he didn't believe it.

"You're not," I said. "Get to the edge."

Zack was still in my canoe. That was the kind of kid he was. He watched from a safe distance.

I'd begged Jackie to let me bring Zack. Jackie was the wildest-looking, wildest-acting man I'd ever seen. He had long, yellow-white hair that he kept pulled back. He was sinewy and muscular, although he never saw the inside of a gym. He had a raw, visceral animal power—electricity, duende—running through him. Any watch he wore stopped running. Jackie's hunt-club buddies were always hanging around his place, usually swigging on a bottle of whiskey. Jackie also grew roses and strung up millions of holiday lights every year and baked cinnamon buns, but still. Imagine being a kid with no mom raised by a man like him. One person never can be two parents. And as a single mother myself, I knew Jackie would welcome a break. And so would Zack.

As soon as we ate, Zack's problems started. He clambered out of the boat and rushed onto a floating island. When he got back, he was pale. He sat down on the thwart, his head hanging.

"Zack?"

"Ma'am?"

"You okay?"

"My stomach hurts and my head hurts. I keep burping up that stuff. It tastes nasty. I'm feeling sick."

"Was that diarrhea?"

"Yes, ma'am." The kid was so damn polite.

"I don't like it that you're starting to get sick now," I said. I glanced at the marker.

"I think I'll be okay." Almost as soon as he said it, he tipped the boat sideways and retched into the black water.

"Oh my god," I said.

"I'm okay," he said.

"At least you're getting it out of your system."

The boys climbed into their boats and we set off. Zack curled up in the canoe. The hot sun was pulsing down. I dipped my bandanna over the side and spread it across his forehead, then covered his arms so he wouldn't get burned. I offered him water. I paddled.

The lamp oil bottle had clearly said Poison. It said in case of ingestion to call the National Poison Hotline. Poison kills. As dangerous as it was to continue, I hated Zack missing the trip. I kept thinking, "Maybe."

At mile #5 I knew I'd made the wrong decision.

The adults huddled. We decided to switch boats so I could take a kayak, which would move faster. I'd pull Zack in a second kayak. My gear would get divided. I'd go back, they'd go on, I'd catch up. The other adults would take care of Silas and Max.

"Silas, I'll be back. It may be after dark. You're not going to worry, are you?"

"I'm not worried," he said.

"Well, please don't worry. I'll be okay. But this is going to take hours, you know that?"

"It's okay," he said. I have never seen my kid intentionally hurt anyone or anything.

I was strong. I could easily pull a boat roped to a second boat where a boy lay, his seat reclined as far as it would go. The drag of his weight was nothing. The first mile was nothing. The second mile was easy. On the third mile I wanted to rest, but Zack was worse. He would lift himself and vomit, then fall back, pale as a tissue.

I pulled right, I pulled left, right, left, right, left, right, left. I jabbed the double-bladed paddle into the water on the right as far to the bow as I could reach, pulled as hard as I could as far to the stern as I could, felt the boat surge, plowing through its own wake, if that's possible. Fetterbush and lizardtail and duck-potato blurred on either side. With the left blade I stabbed forward, using my entire body to torque, pulling hard until the blade was aft. Ahead, a wood duck screamed and flew around a bend. I stabbed right again, left, right, left. We were flying, me with the boy's pack in my boat, the woebegone boy in his plastic boat, sick. When I reached the wood duck, she screamed and fled anew, over and over.

With each stroke I pulled farther from my own son and our friends, but closer to help for Zack. I was paddling upstream on the River Styx, through hot lava, trying to escape a giant arm reaching toward me from the underworld.

I was in a horror movie. The camera was panning. A woman was in focus, churning two boats forward while the landscape streamed by in a blur. The woman was becoming bigger and bigger until she was archetypal. She was a warrior, teeth and claws on strings around her neck, bangles rattling on her wrists. She was transforming into one of the matriarchs pictured on a tarot card. Her torso grew into lioness's, horns sprang from her head, and in her hands she wielded a lightning bolt. She left behind a wake of sparks.

I became something more than I'd ever been as I rowed Zack out of the primordial gunk. As I fought, I also birthed. I birthed Zack and I birthed myself. I birthed too the power of any woman to not be afraid, to not let the fear of death stop her from doing what was needed.

We women forget that most of the time we are warriors fighting for life—to birth it, to protect it, to care for it, to honor it, to continue it. As humans, as women, that's the biggest job we do.

Even as my hips grew tight against the yellow kayak, I was simply a tired woman whose shoulder blades stuck out, whose elbows were sharp, with hair in my mouth, with beads of sweat popping on my brow, strung with three hammocks for carrying babies, one on the left, one on the right, and one on the back. My breasts could not fill fast enough for their hunger. Where my stomach should have been was a big gnawing hole.

I returned through all the places we'd visited that morning, scene upon scene in reverse. In the open prairies the sun beat brutally.

I called Zack's name.

He didn't answer.

I had tied the bow of his boat to my prow with no leeway in the rope, to prevent his kayak listing and encumbering us. I couldn't double back to him.

I called more loudly.

"Ma'am?"

"Drink some water," I said. "You'll dehydrate."

"No, thank you," he said. There was that awful politeness.

"Drink it," I said. I hoped I was doing the right thing.

Then I was off again, a human windmill, around and around—seeming, like a windmill, not to move at all.

Carter Prairie slowed me down, with its shallow water and underbrush of weeds. Our boats scratched through it, and I cursed the Fish and Wildlife Service for not clearing the trails more often. Then my boat stopped short, stuck. I leaped from it and sank knee-deep into layers of ropy, pithy, snaky muck. I began running slow-motion through the battery, dragging the boats behind me, until the water suddenly deepened. I pulled myself back into my boat.

Twenty yards ahead it happened again. Then again. We got into a stretch of weedless water and I paddled more furiously. My arms burned. I was wearing out.

Finally, far ahead I spotted a fisherman in a boat. I was already calling to him. *Do you have a phone? Would you dial Jackie Carter? Here's his number. I have a sick kid, a poisoned kid. Could he meet me at the landing?*

Did you reach Jackie? What did he say? Will he be there?

Zack struggled up and vomited again into the water, and instantly something changed. Color began to seep back into his face. He looked around. "Whew," he said. "I feel better now."

I stopped paddling. "Seriously?"

"I think I'm okay now."

"You scared me back there," I said. I loved the kid, as I think I've said.

"That was nasty."

"We'll be at the landing soon. Your dad's coming."

"I want to go back in," he said.

"Not this time."

Jackie had called his doctor, but Zack was acting like a normal kid, waving to his dad, who was standing outside the open door of his black truck, waiting for us, and now Zack was grinning, paddling behind me. I felt a little foolish. We scraped.

"He may be over it," I said.

Jackie didn't say anything. We got out.

"Get your pack, Zack," I said, stooping to untie the knot that held the boats together.

"You going back in?" Jackie asked me.

"I need to," I said. By then it was almost four o'clock. "But I have to get all the way to Maul Hammock."

"Let me get Zack home," he said. "I can take you partway."

"In your motorboat?"

"What else?" he said. He was like that, mean and kind at the same time.

"I hate it," I said. "I really wanted to take him."

"Shit happens."

"I thought he was going to die, back there," I said.

"Yeah."

The mile to Jackie's house was quick. Zack grabbed a basketball and started playing. I sat on the concrete steps that led into Jackie's house, resting, watching purple martins swooping and swinging, going in and out of their white gourds. Jackie had hung cables low, and in some places the birds were barely above head height.

Zack went to the refrigerator on the porch and came out with a drink. "Now this is a real drink," he said.

I laughed. "I can't believe you didn't smell the lamp oil before it actually got to your mouth."

"I know," he said.

"Or once it was on your tongue, why didn't you taste that it wasn't Gatorade?"

"I gulped it," he said. We both laughed, having beaten the odds of a bad mistake. In the yard Jackie had backed his pickup to a boat trailer. His brother was helping clamp it to the hitch.

"You wanted to be a light," I said to Zack.

"It lit me up," he said.

"I'll stop by in a few days," I said.

At the landing Jackie backed his trailer into the water. He unloaded his boat, stuck my yellow kayak across it, yanked the engine cord.

We started out with a roar, which is how he accomplished most things. I crouched in the front puffed up in my life vest and watched.

I get teary at this point. Jackie's dead now—he died a couple of years ago of what was probably his heart. He came in from bush-hogging to take a nap and never woke up. He died with his work boots by his bed.

When he died, a part of Okefenokee Swamp died.

Some of us meditate in old-growth forests. Some of us watch birds. Some of us gaze out at a beautiful view of a lake. Some of us hunt. But the instinct is the same, I think, to understand that the earth is wild, and that we are of the earth, and also wild. Some of us are willing to feel this more strongly than others. Jackie felt it more than anyone I know, so much that he was feral, living outside the strictures of domestication.

He was one of the most untempered people I've known. He wasn't wild like a pheasant hunter who gathers up bird dogs and rambles in knee boots through soft-edged grassland. He wasn't wild like a hunter shooting a grizzly from a helicopter.

He was the pheasant, he was the grizzly.

I loved him for this. I loved him too because he not only repre-sented my place, the beleaguered land of south Georgia—he was my place, the entirety of it. For years he had to keep one foot in the world of humans, putting up Christmas lights, getting alligators out of backyard ponds, shooting fireworks, baking cinnamon rolls. Now he is wholly the land.

That night as I sank into the cradle of the boat, as the sun low-ered, Jackie wielded the rudder effortlessly. Watching him was like watching an athlete. He was an artist at the wheel. He was always on point, strategizing, maneuvering, navigating the tight mazes, jerking his motor up and down to untangle vegetation from the blades. His calculations were exemplary, even as he swigged Lord

Calvert from a bottle and chased it with water. Of course he wasn't wearing a vest.

He cut the motor when he passed the fisherman. "How do," he said.

"Can't beat it," the man said, his blue ball cap turned backward. "Y'all seen any fish want to ride in my boat?"

Jackie laughed loudly. "We'll be on the lookout."

Then I remembered a moment that morning. Zack wasn't sick yet and we were coasting along.

"That cypress tree," I said, "isn't it beautiful?"

"I bet it's a hundred years old," he said.

"I bet you're right." A minute passed with the two of us eyeballing a gorgeous, many-branching, ballet-dancing cypress tree.

Zack would grow up to be a tall, strong man, devoted husband and father, the kind of guy who takes his kids trick-or-treating one week and then hunting the next. I didn't know that yet. And yet I did. In the quiet moment of a boy and his not-mama sitting quietly looking at a magnificent tree, Zack had pointed off. "That's what's pretty to me," he said. "The way the grass looks there." I knew then that, whatever happened, he would be okay.

Jackie had arrived at a quick series of sharp, wooded bends. He was working the rudder like a dancer. "You're great at this," I said.

"Don't you know I was raised in here?" he said. "I've tore up that corner a few times. I lost a string of fish up here, tipped the boat and the fish got away."

I stayed quiet.

"I'll get you as far as I can," he said. He took another hit of Lord Calvert. "At least to 6. That's halfway. But it's all motorless once you cross the wilderness line."

"Six'd be a huge help," I said.

On the north end of Double Lakes, Jackie pulled up at mile

marker #6. The sun was hanging above a head of moss-strung cypress trees to the west. Turkey vultures circled down toward some conclusion I couldn't see, maybe Cowhorse Island. Jackie moved to unload my boat even as his glided forward.

"Let me help," I said.

"Well, unhook that side," he said, impatient.

I locked my boat against the side of his johnboat and climbed down. Falling in was the kind of thing a puny woman would do. I pushed away and centered my paddle with a clank. "Sorry again, Jackie," I said. "I'll see you in a couple of days."

He shrugged.

"I thank you for bringing me all this way."

He still said nothing.

"No worries," I said. "There's only six miles to go."

"That's still a ride." He turned, revved, roared, and was gone.

For a few minutes I sat in my boat, feeling the wilderness descend. I'd been sad when Zack drank the poison, worried when he began to throw up, panicked as his sickness grew, then sad again. I'd paddled ten hard miles, five of them towing a boat with a boy. Six more lay ahead of me. My folks had arrived at the Maul Hammock shelter by now. The adults would be cooking supper. The boys might be helping, though more likely they'd be playing chess, or fishing, or reading.

Seven hundred square miles is a big wilderness. I had a vision of every alligator in the place rising up out of flower-ridden water, sitting on its tail, and waving. There would be a lot of alligators, some of them as long as fourteen feet, and there would be seventy-four teeth per alligator, each sharpened to a pinpoint. Beyond that bad feng shui there would be black bears, and beyond that, panthers. For these, medicinals do not exist.

Something red flickered at the edge of my vision—male cardinal in a stand of impenetrable titi. Farther away a great egret stalked the shadows. A belted kingfisher came wheeling up the water trail, then careened away, stuttering. In two miles I'd be at Ohio Lake, and in four more, home for the night. There was nothing else to do.

Just at dusk I heard loud bugling. Sandhill cranes, four feet tall, were feeding in a prairie, hollering to each other or maybe hollering to be hollering. They tiptoed through the wet spongy depths of sphagnum, plying it with their long bills. They wore crimson caps atop gray cassocks, perfect feathers laid down perfectly.

Night fell and the moon levitated above the inconceivable and primitive swamp. I paddled by its light, staying within the water trail, spotting trail markers with my light. Large splashes sounded near me, and subversive birds loosed kinky calls. A bellowing began that I had never heard but knew to be the mating calls of alligators.

I went on and on. I refused to rest because the vastness of the swamp and my own smallness scared me. I paddled more quietly. I held my breath.

I passed mile marker #11 close enough that I could touch it with a blade of my paddle. I went on through the lily-mad, mercurial brew. Finally I thought I heard a kid laugh. I saw, across open water, a small glow. I hollered. Hollering is one way swamp folk communicated with each other. They had recognizable hollers, and hollers meant things like "I'm almost home. You can quit your worry." The naturalist Francis Harper, when he visited Okefenokee in the 1910s, recorded some of the hollers, and I'd wound up with a copy of the tape.

Somebody hollered back. It was my own child.

When I reached Maul Hammock, Silas was crouched at the edge of the platform. He was a kindhearted boy who would grow up to be a kindhearted man. I reached out and touched him.

"Hi, sweetie," I said. "I made it."

"I knew you would," he said. "I was waiting on you."

My friends wanted to know all the details. How was Zack and how was I? Was I hungry? They had saved spaghetti. Eat and then tell them everything. First I had to pull myself from the boat and sit on the platform until my knees quit shaking.

The next day we would scrabble through Sapling Prairie and Dinner Pond, then Big Water Lake. We would camp in a shelter there. Somewhere along the way we'd leave wet prairie and enter a fast-moving black creek, the Middle Fork of the Suwannee, one of two rivers that rise from the great swamp, so narrow in places that our paddles swiped the walls of trees. We would come out at Billy's Island, on the other side.

I was pretty much hailed as a hero, but I was no hero. What I'd done was remedial, a defenseless mistake in a grand, heroic, mythic life, and I was damn lucky that a remedy had been available.

One of the most important poses in yoga is a warrior asana, knees bent and arms outstretched, looking forward over the right fingers. It's a pose that requires strength and balance, training for both the physical body and the wisdom body to respond when we are called to action. In this asana the demons of ego, fear, and jealousy can be slain. The pose is also a bowing-down, a recognition of limitations. Gazing past the fingers, we see both near and far.

As my yoga teacher likes to remind me, "You need more than a wish. You need burning desire and fierce determination." When I am in this pose, I know that I am in training, learning to be aware, to not turn a blind eye, to not back down, to not give up. Sometimes the only weapon we have is awareness. Sometimes all we have is a little light that we can shine outward into a big darkness. Sometimes, however, we tap into our superpowers, and then we can transcend and bring about transcendence. Sometimes we shoot flaming arrows.

Most of us, most of our lives, are asked to live small. Most of us quit trying very young to live the bigness we know is possible. Now, no matter what I choose or what is asked of me, I know what I became that long, long night I paddled alone through shamanic darkness in the desolate wilderness just this side of the ultimate wilderness. I have seen the warrior.

GRATITUDE

Thanks to my husband, Raven Zapatismo Waters; son, Silas Ray-Burns; and daughter, Skye Ray-Waters.

I am incredibly lucky to be working with Tom Payton of Trinity University Press, as well as with Burgin Eaves Streetman, Sarah Nawrocki, Bridget McGregor, Steffanie Mortis, and all. Thanks to Christi Stanforth for editing. I thank Pam Houston for choosing the manuscript (anonymously) for the Donald L. Jordan Literary Prize and to Allen Gee, Columbus State University, and Donald L. Jordan for the very generous award.

Writer friends walk with me through a spiraling labyrinth, helping me find a good path and stay on it, and for this I thank Susan Cerulean, Daniel Corrie, Holly Haworth, Joni Tevis, and Amy Wright. I'm grateful for the support of Mary Brown, Ellen Corrie, and Maureen DeVos. The work of Stephen Harrod Buhner has been especially important to me.

A semester-long residency at Hollins University allowed me time to gather this collection, and I deeply appreciate the Susan Jackson Center for Creative Writing, as well as Thorpe Moeckel, Lisa J. Radcliff, Jeanne Larsen, Josh Barkan, and Holly Haworth. I am thankful to other universities that provided residencies over the years, including Florida Gulf Coast University, Green Mountain College, Keene State University, the University of Mississippi, and the University of Montana, as well as Philip Ackerman-Leist, Jill Belsky, Judy Blunt, Laird Christensen, Phil Condon, Peter Corcoran, Debra Earling, Robbie Ethridge, Beth Ann Fennelly, Ann Fisher-Wirth, Tom Franklin, Joan Wylie Hall, the late Barry Hannah, John Harris, Susie Harris, Karen Hurd, Mark C. Long, Steve Siebert, Annick Smith, Robert Stubblefield, Rebecca Todd, Jay Watson, Peter Wirth, and A. James Wohlpart.

My environmental writing professor at the University of Montana, Dr. Hank Harrington (1943–2008), opened me to the beauty of the West and to the power of writing about nature. Hank introduced me to many nature writers—via their works or in person—I would come to cherish. My gratitude to Hank never flags.

I acknowledge the life and work of the late artist Michael Winsor.

The late William Kittredge was my major professor at Montana, and I thank him for teaching me how to write. I would not be a writer without Bill. I thank the inspiring Annick Smith, who brought me into her circle in so many ways and who has been pivotal in my life.

A community of southern nature writers is very important to me, and I thank the late Bill Belleville, Franklin Burroughs, Chris Camuto, Susan Cerulean, Thomas Rain Crowe, Dorinda Dallmeyer, Jan DeBlieu, Ann Fisher-Wirth, Latria Graham, Thomas Hallock, Anna O. Hamilton, Will Harlan, Lola Haskins, Holly Haworth, Sean Hill, Dionne L. Hoskins-Brown, Philip Juras, the late Jim Kilgo, John Lane, Drew Lanham, the late Janet Lembke, Brent Martin, Rose McLarney, Jim Minick, Barbara Ras, Laura-Gray Street, Betsy Teter, Joni Tevis, Rick Van Noy, the late Melissa Walker, and Amy Wright, as well as all writers focused on nature and the South not listed here.

I've worked with editors of great energy and attention on the published pieces, and they include Emerson "Chip" Blake, Paul Bogard, Simmons Buntin, Florence Caplow, Susan A. Cohen, Elizabeth Dodd, Robin Patten, Jennifer Sahn, Caroline Stephens, Melissa Wardlow, and Leia Penina Wilson. I thank them.

The poet Jack Gilbert introduced me to the idea of duende. John Huie, then director of the Environmental Leadership Center at Warren Wilson College, funded my trip to Belize—I appreciate all the ways he helped me. Thanks to the late Joaquín Alvarado García and the National Park Service of Costa Rica, as well as to Breeze VerDant for taking care of Silas while I was there. For the Alaska trip I am deeply grateful to Carolyn Servid and the late Doric Mechau of the Island Institute, to the late Richard Nelson, to Hank Lentfer, and also to Anya Maier and Liz McKenzie. Thanks to Mick Womersley for guiding me to the red rock desert and to Irwin Friedman for introducing me to the Slave Canal. The Helena National Forest and the Holter Museum of Art sponsored the Artist*Forest*Community Residency in the Elkhorn Mountains. I'd like to acknowledge Tom and Kay Amsler, Craig Barrow, Rick Bass, Madison Smartt Bell, Wendell Berry, David Bottoms, Roger W. Bowen, Brian Brown, Dave Brown, Norine Cardea, the late Jackie Carter, Zachary Carter, Jeff Chanton, Stephen Corey, Albert and Leeann Culbreath, John T. Edge, Peter Forbes, Peggy Galis, Toby Graham, Neill Herring, Linda Hogan, Francine Jarriel, Larry Kopchak, Angela Faye Martin, Mike McCall, Laura McCarty, Bill McKibben, Joette and Laz and Julian and Nya Mendez, Steve Mesimer, James Murdock, Susan Murphy, Gary Nabhan, Leandra M. Nessel, Stephanie Packard-Hughes, Cherie and Danny Perez, the late Patrick Pritchard, Carlin J. Ray, Dell Ray, the late Franklin Ray, Lee Ada Ray, Rita Carter Ray, the late Stephen Ray, Sarah Robertson, Pattiann Rogers, Sarah Ross, Scott Russell Sanders, the late Serena Satori, Sean Sexton, Sam Stoloff, Helen Whybrow, and so many, many more.

I thank those who read my work, for the time each person gives it, for sharing copies of books and essays, and for recommending the work to others. My greatest desire is to enliven our culture, cultivating and spreading ideas about a world beyond violence and destruction, a wild and inclusive world, a world that is at our fingertips; and to offer the possibility of transformation. I thank those who keep their hearts open to all of life.

CREDITS

"Bird-Men of Belize" appeared as "Belize: A Record of Life" in *Heartstone*, Spring 2003.

"Exaltation of Elk" appeared in *Wildbranch: An Anthology of Nature, Environmental, and Place-Based Writing*, ed. Florence Caplow and Susan A. Cohen (Salt Lake City: University of Utah Press, 2010).

"I Have Seen the Warrior" appeared in *Terrain.org*, November 9, 2020.

"In the Elkhorns" appeared as "On Walking the Hard Miles," in *Camas Magazine*, Fall 2005.

"Landfall" appeared in *Unspoiled: Writers Speak for Florida's Coast*, ed. Susan Cerulean, Janisse Ray, and A. James Wohlpart (Tallahassee: Red Hills Writers Project, 2010).

"Night Life" appeared as "Against Eternal Day," in *Let There Be Night: Testimony on Behalf of the Dark*, ed. Paul Bogard (Reno: University of Nevada Press, 2008).

"Opening the Big W" appeared as "Up against Open" in *The Purcell Suite: Upholding the Wild*, ed. K. Linda Kivi (New Denver, BC: Maa Press, 2007).

"One Meal" appeared in *Camas Magazine*, Summer 2014.

"Okefenokee Swamp, Georgia, Fall" appeared as "Weaving the World," in *Audubon Magazine*, January 2002. It was set to music by composer Pamela J. Marshall and performed as "Weaving the World" by the Assabet Valley Mastersingers on March 19, 2006, in Southborough, Massachusetts.

Janisse Ray is a naturalist and activist and the author of seven books of nonfiction and poetry, including *The Seed Underground: A Growing Revolution to Save Food, Drifting into Darien: A Personal and Natural History of the Altamaha River,* and *Ecology of a Cracker Childhood,* which won the American Book Award. Her work has appeared widely in magazines and journals, and she is the recipient of a Pushcart Prize, the Nautilus Book Award, and numerous other honors. Ray lives on an organic farm inland from Savannah, Georgia.